Springer Proceedings in Energy

Series Editors

The series Springer Proceedings in Energy covers a broad range of multidisciplinary subjects in those research fields closely related to present and future forms of energy as a resource for human societies. Typically based on material presented at conferences, workshops and similar scientific meetings, volumes published in this series will constitute comprehensive state-of-the-art references on energy-related science and technology studies. The subjects of these conferences will fall typically within these broad categories:

- Energy Efficiency
- Fossil Fuels
- Nuclear Energy
- Policy, Economics, Management & Transport
- Renewable and Green Energy
- Systems, Storage and Harvesting
- Materials for Energy

eBook Volumes in the Springer Proceedings in Energy will be available online in the world's most extensive eBook collection, as part of the Springer Energy eBook Collection. To submit a proposal or for further inquiries, please contact the Springer Editor in your region:

Kamiya Khatter (India)
Email: kamiya.khatter@springer.com
Loyola D'Silva (All other countries)
Email: loyola.dsilva@springer.com

Dave Renné · Carly Rixham · Lauren Reddington
Editors

Proceedings of the 52nd American Solar Energy Society National Solar Conference 2023

Transforming the Energy Landscape for All

Springer

Editors
Dave Renné
Dave Renne Renewables, LLC
Boulder, CO, USA

Carly Rixham
American Solar Energy Society
Boulder, CO, USA

Lauren Reddington
American Solar Energy Society
Boulder, CO, USA

ISSN 2352-2534 ISSN 2352-2542 (electronic)
Springer Proceedings in Energy
ISBN 978-3-031-39146-0 ISBN 978-3-031-39147-7 (eBook)
https://doi.org/10.1007/978-3-031-39147-7

This Springer imprint is published by the registered company Springer Nature Switzerland AG
The registered company address is: Gewerbestrasse 11, 6330 Cham, Switzerland

Preface

The American Solar Energy Society is pleased to present this book of Proceedings from its 52nd Annual National Solar Conference, SOLAR 2023, held on the University of Colorado Boulder campus on August 8–11, 2023. The conference theme was "Transforming the Energy Landscape for All." A broad range of renewable energy professionals, including researchers, architects, engineers, solar practitioners, economists, energy analysts, policymakers, and solar enthusiasts from the general public, came together to participate in the exchange of ideas provided by the conference platform. Conference participants shared their knowledge and experiences and learned about progress and best practices toward an effective and more rapid transformation of our global energy system, a system that meets the challenges of climate change, energy access, security, equity, and economic prosperity. Participants who presented their work at the conference authored the papers in this book, and these papers are organized around the five conference tracks, which addressed the major technical, social, and educational issues apparent in achieving a transformation to 100% renewable energy.

Dave Renné
Carly Rixham
Lauren Reddington

Organization

Proceedings Committee Chairs

Reddington, Lauren
Renné, Dave
Rixham, Carly

Proceedings Review Committee

Coleman, Debra	Sun Plans, Mobile, AL, USA
Ferrera, Summer	Sandia National Laboratories, Albuquerque, NM, USA
Krakauer, Nir	The City College of New York, New York, NY, USA
Michaud, Gilbert	Loyola University, Chicago, IL, USA
Middleton, Paulette	Panorama Pathways, Boulder, CO, USA
Sengupta, Manajit	National Renewable Energy Laboratory (NREL), Golden, CO, USA
Simon, Daniel	3D Solar, Inc., Rocklin, CA, USA
Stromberg, Rich	Alaska Center for Energy and Power, Fairbanks, AK, USA
Vandermark, Henry	Solar-Wave Energy, Cambridge, MA, USA
Wang, Julian	The Pennsylvania State University, University Park, PA, USA

Contents

Decarbonizing the Building Sector

SolarDecomp: A Web App for Decomposing Solar Data for Spectrally Selective Building Simulation

Chenshun Chen(✉), Qiuhua Duan, Yanxiao Feng, and Julian Wang(✉)

Department of Architectural Engineering, Penn State University, State College, USA
{cjc7032,jqw5965,cjc7032,jqw5965}@psu.edu

Abstract. Solar radiation plays an important role in solar architecture design. It not only determines the optical regime of the building envelope, but also influences the heating and cooling loads of the building. To understand the impacts of solar radiation on building thermal and lighting performance, computational analysis is essential. However, conventional modeling tools only take broadband solar radiation into consideration, which limits the modeling accuracy since window glazing is spectrally dependent. In this sense, we developed a new solar decomposing tool that can separate major solar irradiation components such as VIS, NIR in GHI, DNI and DHI, by using machine learning algorithms such as extreme boosting regression tree (*XGBoost*). The predictors are meteorological parameters that already exist or can be easily derived from traditional weather files (such as TMY3). Model performance is validated by NREL Solar Radiation Research Laboratory's spectral solar dataset. Integrating these decomposing models, a web app (SolarDecomp) has been developed to enable researchers and designers import solar spectral data into existing building simulation programs for specific simulations in terms of spectrally selective design components.

Keywords: Narrowband solar irradiation · Solar architecture design and simulation · TMY weather files · Spectral solar dataset · Solar architecture

1 Introduction

Solar radiation plays a crucial role in solar architectural design, which strives to optimize the use of the sun's energy for the creation of energy-efficient structures. Several computational tools are employed to examine the thermal and optical attributes of such designs. Typically, weather data for an entire year is loaded into a simulation program to model conditions like a building's daylight environment and energy consumption. A comprehensive solar-radiation dataset is indispensable during this process, ensuring reliable forecasting for solar initiatives. The consistency of the solar resource, as reflected in historical solar data, coupled with the accuracy of the dataset, significantly impacts the accuracy of the estimates [1].

Several comprehensive datasets are readily accessible: 1) The National Solar Radiation Data Base (NSRDB), a creation of the National Renewable Energy Laboratory

D. Renné et al. (Eds.): ASES SOLAR 2023, SPE, pp. 3–12, 2023.
https://doi.org/10.1007/978-3-031-39147-7_1

(NREL) and Sandia National Laboratory. 2) The Surface Meteorology and Solar Energy (SSE) by NASA, offering global coverage of monthly averages, annual data, and data every three hours from 1983–2005. 3) SolarGIS, which merges ground observations, satellite data, and an atmospheric patterns database to deliver highly accurate global solar radiation data. The Typical Meteorological Year (TMY) data files were derived from these datasets. Each TMY file incorporates a complete year's data, created from 12 representative months chosen from the database's total years. The TMY data includes three key solar components: Global Horizontal Irradiance (GHI), Direct Normal Irradiance (DNI), and Diffuse Horizontal Irradiance (DHI). These are crucial for solar analysis or energy simulation. Importantly, all three types of solar irradiance (GHI, DNI, DHI) in the typical weather data files are broadband, signifying the total quantum of Ultraviolet (UV), Visible Light (VIS), and Near-Infrared Radiation (NIR), which are the main constituents of the solar spectrum.

In recent times, growing evidence from the building and solar simulation has underscored the importance of partitioning broadband solar radiation into narrowband or even design and analytics focused on specific spectrum. For instance, among these three solar components, solar VIS radiation is advantageous for conserving electrical lighting and enhancing indoor health due to its ability to regulate circadian rhythm [2]. However, it may also have adverse effects, leading to visual problems such as glare. While solar NIR transmission into the building could aid in reducing the total heating load in cold climates, which is undesirable in hot climates. Recent studies also show that solar VIS and NIR that penetrate through window have different impacts on thermal comfort near window zones [3]. Regarding the design of solar devices or systems, with the advent of spectral selective and nano-structured thin films for energy-efficient envelopes, the independent modulation of solar radiation in different spectral bands has become increasingly viable as a practical solution for improving building energy savings and maintaining indoor visual comfort. The example studies include reversible photothermal windows [4], silver nanorods-embedded smart windows [5], and other light/heat splitting materials [6–9].

However, the full potential of the materials/structures mentioned earlier in building envelopes may not be comprehended thoroughly if corresponding simulations only factor in broadband solar radiation. Even worse, it may lead to misleading or even erroneous outcomes. Traditional weather files usually do not contain the spectral distribution or narrowband details of solar irradiation due to the complexity and high cost associated with spectral data measurement. To bridge this research gap, it is necessary to develop a reconstruction algorithm that is able to decompose solar visible and infrared irradiance from broadband solar radiation and weather data for building simulations. The solar radiation data used for such simulations include GHI, DHI, and DNI, so all these three solar components need to be decomposed to narrowband data. In our previous work, we have already developed predicting models using the machine learning algorithms such as CART for decomposing VIS and NIR components in GHI, DNI, and DHI [10]. ***This work presents our preliminary development of a data conversion portal*** built upon these predicting models so that broadband solar irradiance inside traditional TMY datasets can be replaced by individual solar components (i.e., VIS and NIR components in GHI, DNI, and DHI), then these modified TMY files can be used for spectrum-related building simulations.

2 Methodology

Firstly, we built two estimation models for decomposing solar components (VIS and NIR) from the broadband GHI and DNI, respectively. The GHI decomposing model and the framework of the data portal have been presented in prior work [10]. Based on these two elements (GHI and DNI), it is possible to calculate the VIS and NIR components in DHI via the transposition equation. Subsequently, one can import their original weather files (such as TMY), and then export processed weather files with individual solar components in replace of broadband GHI, DNI, and DHI information.

In this study, all the datasets were obtained from the BMS database of the NREL Solar Radiation Research Laboratory. All the datasets were based in Golden, Colorado with a latitude 39.742° N, a longitude of 105.18° W, and an elevation of 1828.8 m AMSL [12]. In general, two major datasets were curated: weather datasets (including broadband solar irradiance and other typical meteorological data) and spectral solar irradiance data from multiple sources.

1) Weather Dataset

This dataset includes three components. The major component was based on hourly meteorological measurements (HMM) and included the most independent variables for modeling, such as GHI, DNI, DHI, cloud coverage, dry-bulb temperature, dewpoint, relative humidity, and wind speed. The HMM is hourly data by averaging the value of all measurements taken from the 1-h interval and was collected from Jan 1, 2016, to December 31, 2019. In addition, based on the prior studies, both aerosol optical depth and precipitable water vapor parameters are often found in typical weather stations' data collection and are important to the solar spectra, while they are not included in HMM datasets. As such, in this work, we obtained these two atmospheric data from the NREL BMS AOD and PWV (GPS-based PWV) database. As a result, the curated weather dataset has identical variables and formats to the TMY weather file that has been widely applied to solar radiation and building energy performance simulations. All these weather data including broadband solar irradiance were used as the predicting variable candidates in this work.

2) Spectral Solar Dataset

The corresponding spectral solar components (VIS and NIR irradiance) in GHI were extracted from the outdoor solar spectra data (WISER), and the components in DNI were extracted from the outdoor solar spectra data (PGS-100). The WISER dataset was measured by EKO WISER spectroradiometer MS-710, MS-711, and MS-712 from 2016 to 2019. The MS-710, MS-711, and MS-712 instruments have 4nm, 5nm, and 6.5nm spectral bandwidth respectively. Their measurement range is 300 nm–1100 nm, 300 nm–1100 nm, and 900 nm–1700 nm, respectively [13]. The WISER dataset has a higher resolution measurement for both wavelengths and time intervals (typically 5 min, but occasionally 1 min). The hourly spectrum data were calculated by averaging the 5-min interval data for each hour, The average value of all measured points each hour is defined as the value for the time-stamp at the end of the 1-h interval. The PGS-100 dataset was measured through a LICOR LI-1800 spectroradiometer. The instrument has 3.6 nm spectral bandwidth and the usable spectral range of the instrument is 350 to 1050 nm. Data was taken at approximately 0.7 nm intervals (slightly variable, differs for each serial number) every 5 min [32].

3 Results and Discussion

3.1 VIS and NIR Components in GHI and DNI

In our previous work, we demonstrated the feasibility of decomposing broadband GHI into VIS and NIR components by using the CART algorithm, a simple logistic diagram was developed to cluster VIS and NIR into different fractions of GHI, based on meteorological parameters. The regression tree models for VIS/GHI and NIR/GHI were shown in Fig. 1.

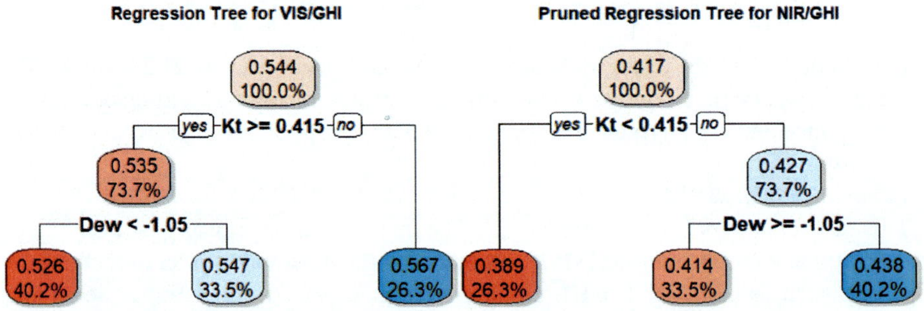

Fig. 1. Regression trees for VIS/GHI (left) and NIR/GHI (right)

Table 1. Model comparison for VIS, NIR in DNI

Model	VIS			NIR		
	RMSE	MAE	R^2	RMSE	MAE	R^2
Cost complexity tree	25.1084	12.9560	0.9650	26.1609	16.9067	0.9791
M5' tree	19.9164	9.6899	0.9780	21.3417	10.6917	0.9861
Random forest	19.1448	9.3176	0.9797	19.9415	9.7087	0.9879
XGBoost	18.2803	7.9894	0.9814	18.3894	8.0108	0.9897

In this study, the general process of predicting VIS and NIR from broadband DNI is shown as an example illustrating the steps of constructing the embedded machine learning algorithm in our SolarDecomp tool. Firstly, by using *sklearn, m5py*, and *XGBoost* packages in Python, we built various regression tree models, including Cost-complexity tree, M5' tree, Random forest and Extreme boosting tree. The entire dataset D was split into a training dataset (80% of D) and a testing dataset (20% of D). Different models were tuned by using cross-validation to find the best training parameters so that each individual model was optimized to fit the training set. After the parameter tuning process, a performance evaluation was conducted to find the best model. As shown in Table 1, XGBoost was found as the best-fit model for predicting both VIS and NIR components in DNI, as it provided the lowest RMSE, MAE, and R^2 results when fitting the testing datasets. Tuned parameters for this optimized *XGBoost* model are shown in Table 2.

Table 2. XGBoost Parameter Tuning

Parameter	Model type	
	VIS	NIR
max_depth	6	4
min_child_weight	5	4
gamma	0	0.4
subsample	0.8	1.0
colsample_bytree	0.8	1.0
reg_alpha	100	100

3.2 VIS and NIR Components in DHI

Solar transposition rule has been frequently used to describe the mathematical relationship among GHI, DHI, and DNI on a horizontal surface: GHI is the sum of incident DHI plus the DNI projected onto the horizontal surface.

$$GHI = DNI \times \cos(sza) + DHI \tag{1}$$

Theoretically, one can use this equation to compute VIS and NIR components in DHI if one takes VIS and NIR components in DNI and GHI as input. However, it is unknown whether this relationship is still valid for each spectral solar component. To verify the transposition model, two datasets (collected between May 12th and June 1st 2021) in the BMS database of the NREL Solar Radiation Research Laboratory were selected. The measured spectral DHI data were obtained by EKO MS-711 and MS-712 sensors, and then VIS and NIR of DHI were prepared. On the other hand, the RG 780 dataset (Schott Glass filter with a cut-on wavelength at 780nm) was used to extract VIS and NIR components from GHI and DNI, which were then input into the transposition equation to compute the VIS and NIR components of DHI. As such, both measured and calculated VIS and NIR irradiance of DHI were obtained and then compared.

Figure 2 shows the comparison results of VIS in DHI, the comparison of NIR in DHI has similar results. The average relative error between the measured VIS and NIR data and the data calculated by the transposition model is 7.62% and 6.89% for VIS and NIR, respectively. This comparison demonstrates that the diffuse, direct, and global spectral solar irradiance values are also complying with the relationship defined by the transposition model. After the solar transposition model was validated in the broadband solar spectrum, VIS and NIR components in DHI can be calculated by using previously predicted VIS and NIR in GHI and DNI as inputs to the solar transposition model. In general, NIR occupies a small portion of DHI since the most diffused solar irradiance is from the sky and tends to be in short wavelengths. There are potential errors with this prediction and computation method as the calculated spectral component of DHI might accumulate prediction errors from both DNI and GHI. As such, in the actual application, a set of calibrations will be processed before the final spectral data of solar are reported to users.

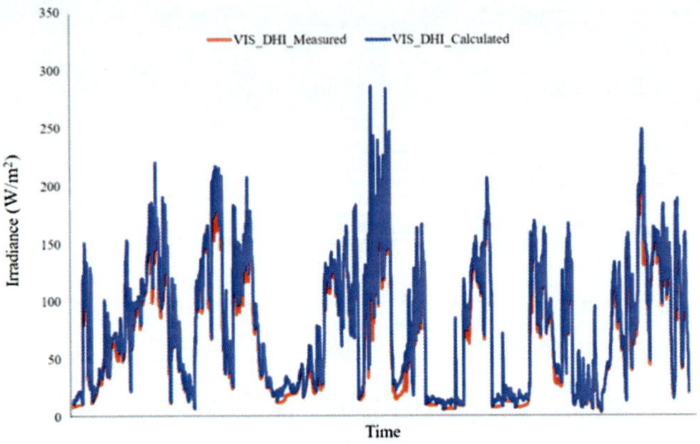

Fig. 2. Comparison of measured/calculated VIS results in DHI

4 Web Portal Development of Solar Decomposing in Solar Architecture Design

The GHI, DNI, and DHI solar decomposing models were integrated into an executable web app, SolarDecomp, via Python *streamlit* package. After a user chooses their preferable solar component, they will be asked to upload a weather file (e.g., TMY file) and the location of that weather file, then SolarDecomp would automatically execute previously trained XGBoost model and generate a different weather file in which the original broadband GHI, DNI, and DHI are replaced with the narrowband solar components (VIS or NIR). The final outputs would be new weather files labeled TMY_VIS and TMY_NIR. Note that the input weather file should be pre-processed to ensure that the input file is in.csv format (it is recommended to use EnergyPlus™ Weather Converter to convert the original.epw file to.csv file), all features used for predicting exists (except for newly added parameters, such as K_b and *sza*, which will be automatically computed based on the existing parameters) and all features' names and units should be in the same format. Table 3 summarizes all the required input variables and their name formats. General instruction of using SolarDecomp is shown in Fig. 3.

It's also worth mentioning that the predicting values will be automatically calibrated before exporting the final output files. The calibration logic is shown in Fig. 4. While feeding the model with weather files, if any features were missing or their numbers were meaningless (e.g., some stations do not provide Albedo, liquid precipitation depth, snow depth, AOD, or PWV data), a simplified model will be used to replace of the full model. This simplified model could only use part of the features to make predictions: for example, GHI, DNI, DHI, SKC, Opqcld, Dry, Dew, RH, AM, I_0, Kb, and Tcld. Normally, the simplified model can be used at most of the locations, but this generalizability is a consequence of losing accuracy (e.g., the simplified model has a testing RMSE = 20.6103, MAE = 10.2091, R^2 = 0.9764 for VIS/DNI decomposing).

With such new weather files, designers and researchers are now able to integrate separable solar irradiance when simulating a building solar environment by using simulation tools such as ClimateStudio, EnergyPlus, etc. Figure 5 demonstrates a simple

Table 3. Model required input variables and their unified formats

Feature	Name format	Units format	Source
Global Horizontal Irradiance	GHI	W/m^2	TMY
Direct Normal Irradiance	DNI	W/m^2	TMY
Diffuse Horizontal Irradiance	DHI	W/m^2	TMY
Total sky cover	SKC	.1	TMY
Opaque sky cover	Opqcld	.1	TMY
Dry-bulb temperature	Dry	°C	TMY
Dew-point temperature	Dew	°C	TMY
Relative Humidity	RH	%	TMY
Atmospheric pressure	Pressure	Mbar	TMY
Zenith angle	Zenith	°	Calculation
Air mass	AM		Calculation
Extraterrestrial Direct Normal Radiation	I0	W/m^2	Calculation
Normal clearness index	Kb		Calculation
Cloud transmittance	Tcld		Calculation
Azimuth angle	Azimuth	°	Calculation
Wind direction	Wdr	° from N	TMY
Wind speed	Wspd	m/s	TMY
Albedo	Albedo	.01	TMY
Liquid precipitation depth	Precip	mm	TMY
Snow depth	Snow	cm	TMY
Aerosols Optical Depth	AOD	.001	TMY
Precipitable Water Vapor	PWV	mm	TMY

example of investigating the surface irradiance in a high-dense urban area. The urban model contains all-glass buildings, on which Double Low-E windows were applied. These Low-E windows had relatively high VIS transmittance (51%) but extremely low NIR transmittance (0.1%) (glazing materials' spectral properties modeled in LBNL Optics software, and averaged across the spectrum). The original weather file (Philadelphia_International_Ap:: 724080:: TMY3) was modified by the data conversion tool and then imported to simulation software (ClimateStudio in Rhino). With the original weather file as the input, ground solar irradiance in July was around 138 kWh/m^2, however, after separating the simulation into VIS and NIR parts, the total summation of VIS and NIR irradiance was around 176 kWh/m^2, much larger than the original simulation. Although this demonstration is simple and inaccurate, it still reveals the necessity of separating solar components for building solar simulations, especially when considering spectral-selective windows.

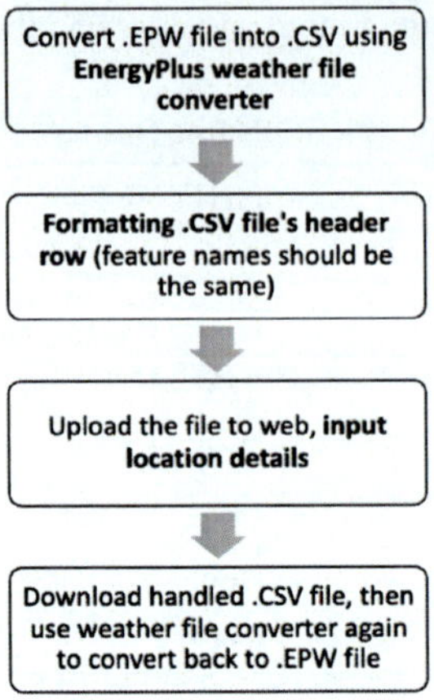

Fig. 3. Instruction for converting traditional weather file to VIS/NIR weather file by using SolarDecomp

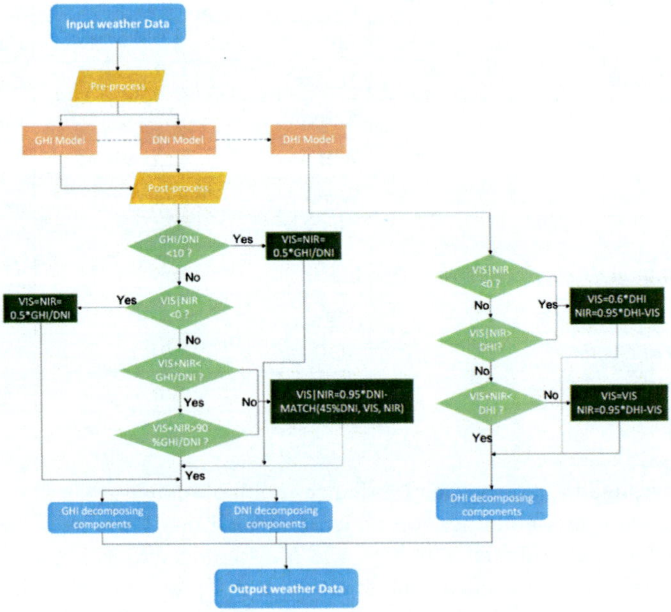

Fig. 4. Logistic diagram of model calibration

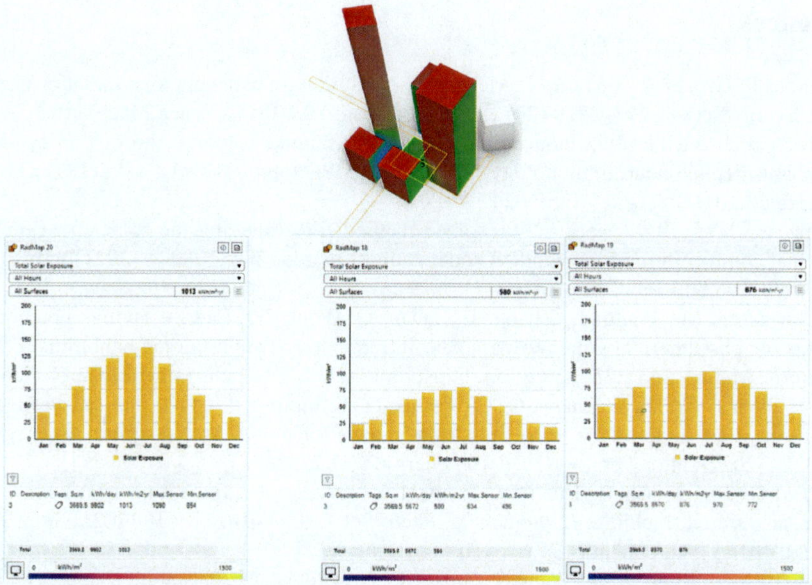

Fig. 5. Urban solar irradiance simulation demo: Bottom left: Ground solar irradiance with original weather file; Bottom right: VIS and NIR solar irradiance with modified weather file

5 Conclusion

In conclusion, we have successfully developed an executable web app, SolarDecomp, leveraging the Python streamlit package, which integrates models for decomposing GHI, DNI, and DHI solar data. Users can upload a weather file and its location, and the app will use the pre-trained XGBoost model to generate a modified weather file with distinct narrowband solar components (VIS or NIR). The result is new weather files, TMY_VIS and TMY_NIR. The input file should be pre-processed into a.csv format, with all necessary predictive features present and correctly formatted. This research highlights the effective use of machine learning to convert broadband solar data into separate components, utilizing readily available meteorological data. Despite these advancements, there remain research gaps such as the need for solar analyses on tilted and vertical surfaces and model validation under different climate conditions, the generalizability of trained model by using solar spectral information from different locations, which will be addressed in future work.

Acknowledgements. This project is supported by the NSF award: # 2001207: CAREER: Understanding the Thermal and Optical Behaviors of the Near Infrared (NIR)-Selective Dynamic Glazing Structures.

References

1. Vignola, F., Grover, C., Lemon, N., McMahan, A.: Building a bankable solar radiation dataset. Sol. Energy **86**(8), 2218–2229 (2012). https://doi.org/10.1016/j.solener.2012.05.013
2. Roberts, J.E.: Visible light induced changes in the immune response through an eye-brain mechanism (photoneuroimmunology). J. Photochem. Photobiol. B **29**(1), 3–15 (1995). https://doi.org/10.1016/1011-1344(95)90241-4
3. Wang, N., Wang, J.: A spectrally-resolved method for evaluating the solar effect on user thermal comfort in the near-window zone. Build. Environ. **202**, 108044 (2021). https://doi.org/10.1016/j.buildenv.2021.108044
4. Anwar Jahid, M., Wang, J., Zhang, E., Duan, Q., Feng, Y.: Energy savings potential of reversible photothermal windows with near infrared-selective plasmonic nanofilms. Energy Convers. Manag. **263**, 115705 (2022). https://doi.org/10.1016/j.enconman.2022.115705
5. Pu, J., Shen, C., Yang, S., Zhang, C., Chwieduk, D., Kalogirou, S.A.: Feasibility investigation on using silver nanorods in energy saving windows for light/heat decoupling. Energy **245**, 123289 (2022). https://doi.org/10.1016/j.energy.2022.123289
6. Du, W.-C., Xie, J., Xia, L., Liu, Y.-J., Yang, H.-W., Zhang, Y.: Study of new solar film based on near-infrared shielding. J. Photochem. Photobiol. Chem. **418**, 113410 (2021). https://doi.org/10.1016/j.jphotochem.2021.113410
7. Besteiro, L.V., Kong, X.-T., Wang, Z., Rosei, F., Govorov, A.O.: Plasmonic glasses and films based on alternative inexpensive materials for blocking infrared radiation. Nano Lett. **18**(5), 3147–3156 (2018). https://doi.org/10.1021/acs.nanolett.8b00764
8. Gao, Q., Wu, X., Huang, T.: Novel energy efficient window coatings based on In doped CuS nanocrystals with enhanced NIR shielding performance. Sol. Energy **220**, 1–7 (2021). https://doi.org/10.1016/j.solener.2021.02.045
9. Xu, Q., et al.: Cs0.33WO3 as a high-performance transparent solar radiation shielding material for windows. J. Appl. Phys. **124**(19), 193102 (2018). https://doi.org/10.1063/1.5050041
10. Duan, Q., Feng, Y., Wang, J.: Clustering of visible and infrared solar irradiance for solar architecture design and analysis. Renew. Energy **165**, 668–677 (2021). https://doi.org/10.1016/j.renene.2020.11.080
11. Myers, D.R.: Solar Radiation: Practical Modeling for Renewable Energy Applications, 1st ed. CRC Press (2017). https://doi.org/10.1201/b13898
12. Achleitner, S., Kamthe, A., Liu, T., Cerpa, A.E.: SIPS: Solar irradiance prediction system. In: IPSN-2014 Proceedings of the 13th International Symposium on Information Processing in Sensor Networks, pp. 225–236, April 2014. https://doi.org/10.1109/IPSN.2014.6846755
13. MIDC: SRRL BMS Instruments. https://midcdmz.nrel.gov/srrl_bms/instruments.html. Accessed 16 Dec 2022

Evaluation of Six Greenhouse Covering Materials for Energy Performance

Enhe Zhang, Md Anwar Jahid, and Julian Wang(✉)

Department of Architectural Engineering, Penn State University, University Park, PA 16802, USA

{ekz5068,mmj5702,julian.wang}@psu.edu

Abstract. Greenhouse technology plays a vital role in meeting the global demand for food production by enabling year-round crop cultivation in more than 55 countries. Selecting suitable cover materials is crucial for greenhouse systems as they significantly impact the transmission of Photosynthetically Active Radiation (PAR), heat retention, and total energy efficiency. This study investigates six greenhouse coverings: clear glass, clear glass with UV coatings, double-pane glass, double-layer polyethylene (PE), double-layer polycarbonate (PC), and double-pane glass with UV coatings, focusing on their optical and thermal properties. The optical properties examined include spectral transmittance, absorptance, reflectance, and solar heat gain coefficient (SHGC), which directly influence the amount of solar radiation and PAR entering the greenhouse. The SHGC and U-factor are key parameters for assessing the energy efficiency of these cover materials. The study employed the LBNL Window and Optics software to calculate the SHGC and U-factor for each covering configuration. An EnergyPlus model evaluated comprehensive annual energy consumption for each covering configuration. The results revealed that double-layer PE exhibited remarkable energy savings of 26.9% in heating energy usage compared to single-pane glazing. Furthermore, double-layer PE demonstrated a 10.1% reduction in total energy consumption compared to single-pane glazing. Interestingly, the study also found that the impact of PAR transmittance on interior lighting was relatively minimal.

1 Introduction

Greenhouse technology has been employed for centuries to extend the growing season, enhance crop yield and quality, and mitigate pest infestations. As the global demand for food continues to rise, greenhouse systems have gained prominence in over 55 countries, enabling year-round crop production [1]. An essential component of these systems is the meticulous selection of appropriate covering materials, crucial in regulating the transmission of Photosynthetically Active Radiation (PAR), heat retention, and total operational energy use efficiency. Clear glass, the predominant greenhouse cover material, exhibits high solar and PAR transmittance and significant solar reflectance and emissivity. UV adhesive film is often applied to clear glass to protect the materials, allowing ample solar and PAR transmittance while maintaining a high solar heat gain coefficient [2, 3]. Additionally, lightweight and flexible plastic materials such as polyethylene (PE) and polycarbonate (PC) are commonly used as greenhouse coverings [4, 5]. Thus, this study focuses on four representative materials: clear glass, UV-coated glass, PE, and PC.

D. Renné et al. (Eds.): ASES SOLAR 2023, SPE, pp. 13–20, 2023.
https://doi.org/10.1007/978-3-031-39147-7_2

The preceding research aims to characterize these representative greenhouse covering constructions, including clear glass, clear glass coated with UV, double-pane glass, double-pane clear glass with UV coating, double-layer PE, and double-layer PC, in terms of their optical and thermal properties. The optical properties encompass spectral transmittance, absorptance, reflectance, and solar heat gain coefficient (SHGC), while the thermal properties considered in this study is U-factor. The optical properties assume particular significance as they influence the amount of solar radiation and PAR that can penetrate the greenhouse structure. Conversely, the SHGC and U-factor serve as key parameters for assessing the energy efficiency of greenhouse covering materials [6–8]. Based on the measurements obtained for each material's optical and thermal properties, SHGC and U-factor values were derived using the LBNL Window and Optics software [9]. The underlying motivation of this study is to gain insights into the thermal and optical distinctions among these materials and their potential impacts on the energy performance of greenhouse operations, encompassing heating, cooling, interior lighting, and total energy consumption. To accomplish this, a simulation model will be constructed to evaluate the six different greenhouse covering comprehensively.

Numerous researchers have dedicated their efforts to investigating the influences of greenhouse covering materials through simulation studies [10–14]. Al-Mahdouri et al. developed an agricultural greenhouse model. They utilized radiative heat transfer analysis to examine the optical properties and thermal behavior of PE, PO, PVC, and Glass cover materials. By employing the Radiative Element Method (REM2), they observed that higher infrared (IR) reflectance resulted in lower surface temperatures for the covering materials [10]. Kavga et al. introduced a comprehensive dynamic greenhouse simulation tool designed to evaluate crop thermal well-being and energy requirements. With a focus on window glazing as a critical component, the study validated the simulation tool TRNSYS by comparing results with EnergyPlus. This validation process accounted for factors such as the U-factor, absorptance, emissivity, and convective heat transfer coefficient settings, resulting in a noteworthy yearly average relative error of 11.1% [12]. Baxevanou et al. conducted an annual numerical evaluation of greenhouse cover materials, specifically investigating PE, EVA, and PVC. Their objective was to determine the longest satisfaction period for tomato crops, considering key factors such as photosynthetically active radiation (PAR) transmittance, temperature, and air velocity. The study used ANSYS Fluent Computational Fluid Dynamics (CFD) software to analyze these cover materials comprehensively [15]. Stefano et al. proposed an innovative greenhouse material based on numerical, optical, thermal, and mechanical approaches. The study focused on a new variant of high-density polyethylene (HDPE) as the cover material. It employed numerical simulations using COMSOL software to assess its optical, thermal, and mechanical characteristics. Remarkably, the research indicated a significant 50% reduction in thermal conductivity compared to the original HDPE material [16]. Despite the significant research efforts dedicated to greenhouse simulations, it is important to note that many studies have overlooked the crucial influence of optical and thermal properties of covering materials on the annual energy performance of greenhouses. These simulations focus primarily on factors such as crop well-being, heating and cooling requirements, and total energy needs. However, the covering materials' specific optical and thermal characteristics, which play a vital role in controlling heat transfer and light transmission, have often been disregarded.

The main goal of this study is to obtain a holistic comprehension of the thermal and optical distinctions among the six different greenhouse covering constructions and their consequent effects on the potential energy performance of greenhouse operations, as well as the indoor conditions crucial for photosynthesis. Additionally, we seek to evaluate their influence on total energy utilization, encompassing heating, cooling, and interior lighting within the greenhouse environment. By thoroughly examining these aspects, we aim to offer valuable insights that can be utilized to optimize greenhouse energy efficiency, minimize energy consumption, and establish conducive environments that promote efficient photosynthesis.

1.1 Methodology

The methodology employed in this study consists of two primary components: (1) the acquisition of optical and radiometric data pertaining to the greenhouse covering materials and the subsequent determination of their solar heat gain coefficient (SHGC), U-factor, and visual transmittance; and (2) the development of a greenhouse model in EnergyPlus, incorporating the diverse covering materials as window surfaces and conducting simulations to assess the energy performance of the greenhouse.

Part 1: Material Measurements and Characterization.

In this segment, we performed measurements of the optical and radiometric properties of the greenhouse covering materials utilizing a spectrophotometer and pyranometer. The materials subjected to analysis were clear glass, polyethylene (PE), polycarbonate (PC), and an ultraviolet (UV) coating. Comprehensive data was collected concerning their transmittance, reflectance, and absorptance. By utilizing advanced software tools (previous work), namely Optics 6 and Window 7.8, the acquired data was subsequently employed to compute the materials' SHGC, U-factor, and PAR transmittance, which were subsequently utilized in the EnergyPlus simulations. Despite maintaining a similar U-factor for the double pane/layer structures, notable differences were observed in terms of PAR transmittance and SHGC.

Part 2: Greenhouse Model and Simulation.

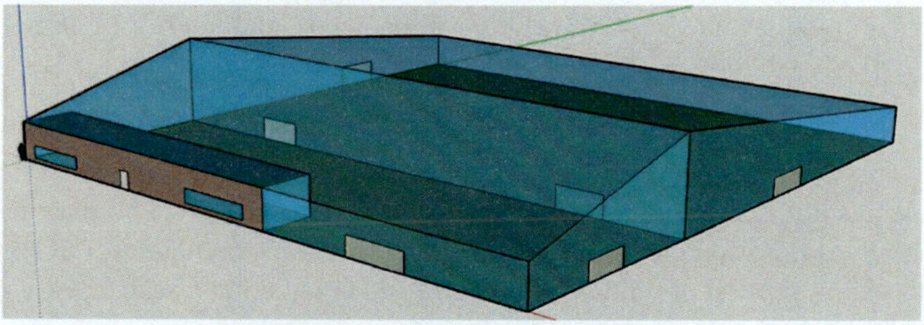

Fig. 1. The greenhouse simulation model.

The DOE Prototype Building Model: Warehouse (non-refrigerated) 90.1 2019, has undergone modifications to align with greenhouse standards and incorporate specific features and controls. An important modification is the incorporation of window glazing on all exterior walls and the roof, aligned with the original structure's dimensions as shown in Fig. 1. Also, further updates for greenhouse operational setting are referred and listed in following [12, 17]. This addition maximizes the utilization of natural daylight. Additionally, the roof has been raised to a 30-degree slope to comply with greenhouse structural requirements and design principles. The daylight control system operates continuously, maintaining a setpoint of 8475 lx and 22 Discomfort Glare Index (DGI). This ensures that artificial lighting is only engaged when natural light levels dip below the predetermined threshold. The lighting system operates for 14 h each day, all year round, including weekends, holidays, and both summer and winter design days. The lighting power is 57.25 W/m^2, providing ample illumination from 7 am to 9 pm (14 h). The HVAC system has been adjusted to maintain a constant heating setpoint of 16 °C and cooling setpoint of 28 °C throughout the day. The zone supply air has been appropriately sized, accounting for a cooling air temperature of 14 °C and a heating air temperature of 50 °C. Heating is achieved through a natural gas fuel-based heating coil, while cooling is accomplished by an air-cooled condenser powered by electricity. The fan system has been optimized by reducing the design pressure rise to 240 Pa, minimizing energy consumption associated with fan operation. To evaluate the impact of the different greenhouse covering materials, we introduced them into the model as window surfaces by adjusting their U-factor, SHGC, PAR-to-solar gain (PSG), and visual transmittance parameters. Subsequently, the EnergyPlus simulation was executed over an entire year, during which the total energy consumption, heating energy, and lighting energy were meticulously recorded for each simulation. By comparing the results obtained from each simulation with those of the baseline model, we were able to discern and quantify the influence of the various covering materials on energy consumption.

2 Results

The results of this study can be divided into two main parts: the analysis of optical and thermal properties of the greenhouse covering materials, and the energy performance evaluation through simulations using EnergyPlus and the greenhouse model. In this section, we will discuss how three key factors, namely PAR transmittance, solar heat gain coefficient (SHGC), and U-factor, influence energy performance in terms of total energy consumption, heating requirements, cooling needs, and interior lighting.

2.1 Optical and Thermal Properties

Table 1 summarizes the greenhouse covering materials, focusing on the PSG (Photosynthetically Active Radiation) value, which represents the ratio of PAR (Photosynthetically Active Radiation) transmittance to SHGC (Solar Heat Gain Coefficient). PSG provides insight into the proportion of photosynthetically active radiation transmitted relative to potential solar heat gain.

As shown in Fig. 2 and Table 1, UV-coated materials exhibit high PSG values of 1.36 and 1.56, indicating superior performance in the photosynthetically active region.

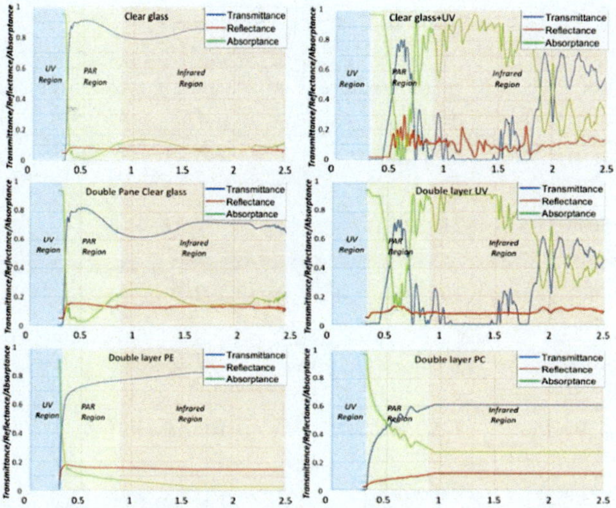

Fig. 2. Optical data, transmittance, reflectance, and absorptance.

Table 1. Optical and thermal properties

	T_{par}	SHGC	PSG	U-factor, W/m^2K
Clear glass	0.902	0.869	1.04	5.879
UV-coated glass	0.636	0.467	1.36	5.858
Double-pane glass	0.797	0.741	1.07	2.903
Double-pane UV	0.572	0.367	1.56	2.944
Double-layer PE	0.783	0.793	0.99	2.911
Double-layer PC	0.489	0.671	0.73	2.877

These materials allow more photosynthetically active radiation to pass through while reducing solar heat gain. Conversely, double-layer PC (polycarbonate) has a lower PSG of 0.73, indicating reduced light transmission and increased heat gain through material absorption. Moreover, the PAR transmittance values demonstrate differences among the materials. Clear glass has the highest PAR transmittance, nearly double that of a double-layer PC. Additionally, the applied UV coating significantly reduces the SHGC of clear glass by 46.2%.

3 Energy Performance

Table 2 provides a comprehensive overview of the total energy and heating values associated with various greenhouse covering materials. Notably, the double-layer PE exhibits a remarkable total energy saving of approximately 10.1%, closely followed by the double-layer PC with a savings of about 7.9% compared to single-pane glass. Furthermore, in

Table 2. Energy performance.

name	Total Energy(GJ)	Heating(natural gas) (GJ)	Cooling(electricity) (GJ)	Interior lighting (GJ)
Single pane Glass	4891.22	2050.39	95.85	2165.55
Single-pane UV	5071.12	2173.62	92.37	2232.06
Double pane Glass	4408.92	1521.63	102.36	2177.62
Double-pane UV	4624.32	1693.74	97.67	2255.57
Double-layer PE	4395.25	1499.16	103.15	2181.55
Double-layer PC	4503.33	1533.00	101.98	2265.59

terms of heating energy, the double-layer PE and double-layer PC demonstrate substantial savings of approximately 26.9% and 25.2%, respectively. This indicates their effectiveness in reducing heating energy consumption, contributing to total energy savings. These findings highlight the potential of using double-layer PE or PC materials to achieve significant energy efficiency improvements in greenhouse operations. In contrast, introducing a UV coating to clear glass results in a modest increase of approximately 3.7% in total energy consumption and 6.0% in heating energy. The presence of the UV coating modifies the optical and thermal properties of the covering material, leading to observed variations in energy consumption.

Tables 1 and 2 also reveal an intriguing relationship between the U-factor and energy consumption. The transition from single pane to double pane structure significantly reduces the U-factor from 5.879 to 2.903 W/m^2K. This general trend of lower U-factor values is associated with reduced total and heating energy consumption. Consequently, the double-pane configuration outperforms single-pane glass in terms of energy efficiency. However, the influence of SHGC and PAR transmission must be considered. Comparing structures with similar U-factor values, such as double pane UV and double pane glass, it is evident that double pane UV exhibits higher total energy consumption by approximately 4.9% and higher heating energy consumption by 11.3%. The lower SHGC value of double pane UV, resulting in reduced solar heat gain, contributes to the higher energy use for heating. This effect is predominantly responsible for influencing total energy consumption. Similarly, despite its lower U-factor, double-layer PC exhibits the highest heating and total energy consumption among similar structures. This can be attributed to its lowest SHGC and PAR transmittance values, which further hinder solar heat gain and contribute to increased energy demand.

Furthermore, the data on PAR transmittance reveals a relationship between PAR transmittance and interior lighting, albeit with a relatively minimal impact. Across all simulations, the introduction of double-layer PC covering material, which exhibits a

46.2% reduction in PAR value compared to clear glass, results in a slight increase of 4.61% in interior lighting usage. This suggests that the decrease in PAR transmittance has a limited effect on the amount of interior lighting required.

4 Conclusion

In conclusion, this study utilized an EnergyPlus Model to assess the energy performance of different greenhouse covering materials. The findings indicate notable variations in total and heating energy consumption, with double-layer PE and PC materials demonstrating high energy efficiency. The U-factor emerged as a crucial factor influencing energy consumption, while the interplay between U-factor, SHGC, and PAR transmission warrants consideration. Additionally, the study revealed that the impact of PAR transmittance on interior lighting was relatively minimal. It is important to acknowledge the limitations of the study, such as the absence of experimental validation. Further research incorporating practical experiments and optimizing greenhouse design is recommended. By identifying energy-efficient materials, this study contributes to sustainable agricultural practices. The results highlight the need to consider the energy performance of greenhouse coverings and support the development of greener practices in greenhouse agriculture.

Acknowledgements. We acknowledge the financial support provided by the USDA Natural Resources Conservation Service NR203A750008G006.

References

1. Sahdev, R.K., Kumar, M., Dhingra, A.K.: A comprehensive review of greenhouse shapes and its applications. Frontiers in Energy **13**(3), 427–438 (2017). https://doi.org/10.1007/s11708-017-0464-8
2. Timmermans, G.H., et al.: Advanced optical materials for sunlight control in greenhouses. Advanced Optical Materials **8**, 2000738 (2020)
3. Ravishankar, E., et al.: Balancing crop production and energy harvesting in organic solar-powered greenhouses. Cell Reports Phys. Sci. **2**, (2021)
4. Choab, N., et al.: Review on greenhouse microclimate and application: Design parameters, thermal modeling and simulation, climate controlling technologies. Sol. Energy **191**, 109–137 (2019)
5. Maraveas, C.: Environmental sustainability of greenhouse covering materials. Sustainability (Switzerland) 11, 6129 (2019)
6. Duan, Q., Zhang, E., Hinkle, L., Wang, J.: Parametric energy simulation methods for solar-NIR selective glazing systems. J. Phys. Conf. Ser. **2069**, 012129 (2021)
7. Zhang, E., Jahid, M.A., Wang, J., Wang, N., Duan, Q.: Experimental investigation of the thermal performance of greenhouse glazings in surface condensation conditions. Appl. Therm. Eng. (2023). Under revision
8. Jahid, M.A., Wang, J., Zhang, E., Duan, Q., Feng., Y.: Energy savings potential of reversible photothermal windows with near infrared-selective plasmonic nanofilms. Energy Convers. Manag. **263**, 115705 (2022)

9. Ida, A.: Multivalent analysis of double-skin envelope dynamic hygrothermal louver system. In: Simulation Series, vol. 50 (2018)
10. Al-Mahdouri, A., Baneshi, M., Gonome, H., Okajima, J., Maruyama, S.: Evaluation of optical properties and thermal performances of different greenhouse covering materials. Sol. Energy **96**, 21–32 (2013)
11. Kavga, A., Souliotis, M., Koumoulos, E.P., Fokaides, P.A., Charitidis, C.A.: Environmental and nanomechanical testing of an alternative polymer nanocomposite greenhouse covering material. Sol. Energy **159**, 1–9 (2018)
12. Baglivo, C., et al.: Complete greenhouse dynamic simulation tool to assess the crop thermal well-being and energy needs. Appl. Therm. Eng. **179**, 115698 (2020). https://doi.org/10.1016/j.applthermaleng.2020.115698
13. Al-Mahdouri, A., Gonome, H., Okajima, J., Maruyama, S.: Theoretical and experimental study of solar thermal performance of different greenhouse cladding materials. Sol. Energy **107**, 314–327 (2014). https://doi.org/10.1016/j.solener.2014.05.006
14. Kim, H.K., Lee, S.Y., Kwon, J.K., Kim, Y.H.: Evaluating the effect of cover materials on greenhouse microclimates and thermal performance. Agronomy **12**, 143 (2022)
15. Baxevanou, C., Fidaros, D., Bartzanas, T., Kittas, C.: Yearly numerical evaluation of greenhouse cover materials. Comput. Electron. Agric. **149**, 54–70 (2018)
16. Sfarra, S., et al.: A proposal of a new material for greenhouses on the basis of numerical, optical, thermal and mechanical approaches. Constr. Build. Mater. **155**, 332–347 (2017)
17. Choab, N., et al.: Effect of greenhouse design parameters on the heating and cooling requirement of greenhouses in moroccan climatic conditions. IEEE Access **9**, 2986–3003 (2021)

Quantifying Solar Light-Induced Thermal Comfort Effects of Architectural Windows

Nan Wang, YanXiao Feng, and Julian Wang(✉)

Pennsylvania State University, University Park, PA 16801, USA
julian.wang@psu.edu

Abstract. Shortwave solar irradiance has an energy distribution at different wavelengths and can impact the indoor thermal comfort of humans by entering through windows and directly affecting human skin. Windows play a crucial role in this process, and the market offers a wide range of products with various technical designs and functions for selection. Thermal comfort effects induced by solar irradiance were simulated and quantified by the spectrally-resolved method, specifically focusing on different glazing options in the International Glazing Database. The thermal comfort levels with different glazing installed were also evaluated under different indoor environments. By comparing the thermal comfort levels achieved with different glazing options, it was observed that the choice of glazing system significantly influenced the solar light-induced thermal comfort levels, consequently affecting the overall indoor thermal comfort of humans in diverse indoor environments. Quantifying the solar light-induced thermal comfort effects of architectural windows can be helpful in selecting appropriate windows for buildings, ultimately improving the thermal comfort of humans and energy savings.

Keywords: Spectrally-resolved Method · Window database · Thermal comfort

1 Introduction

Thermal comfort is a crucial aspect to evaluate the human status and indispensable to maintain the productivity and well-being of human beings indoors. The prediction of the thermal comfort level provides support for evaluating and constructing a well-living and working environment. The predicted mean vote (PMV) model is widely utilized to predict human thermal comfort and serves as a fundamental model for incorporating additional factors and specific evaluation targets in the prediction [1]. The input of the PMV model to predict the thermal comfort level includes air temperature, mean radiant temperature (MRT), air velocity, relative humidity, metabolic rate, and clothing level [1]. Among them, MRT represents the longwave solar effect on human thermal comfort. To incorporate the additional shortwave solar irradiance into the model, the effect of the shortwave solar irradiance was converted into the equivalent longwave effect, acting as a supplementary variable. The variable is referred to as MRT delta [2]. The value of MRT delta depends on the presence of shortwave solar irradiance

D. Renné et al. (Eds.): ASES SOLAR 2023, SPE, pp. 21–26, 2023.
https://doi.org/10.1007/978-3-031-39147-7_3

on human skin indoors. The availability of solar irradiance is determined by several factors, including the solar irradiance that reaches the window, window transmittance, and skin absorptivity. These factors exhibit spectral power distributions, which means the energy of solar irradiance, as well as the properties of window transmittance and skin absorptivity, are distributed unequally across different wavelengths. To account for the spectral characteristics of these three key variables and other factors influencing the availability of solar irradiance, a spectrally-resolved method was proposed [3]. This method aims to enhance the accuracy of predicting human thermal comfort indoors.

The International Glazing Database (IGDB) offers optical data for a range of glazing products [4]. Manufacturers of these glazing products have measured the spectral transmittance and reflectance by a spectrophotometer, and the data has been contributed to the IGDB. Assessing the performance of the glazing products within the database is helpful for gaining a comprehensive understanding of the role of windows and their practical application in relation to the solar light-induced thermal comfort effect.

In this paper, the MRT delta and PMV values of the glazing products available in the database were calculated and compared. Additionally, the MRT delta values corresponding to the comfort PMV value were calculated under various indoor environments to investigate the combined impact of multiple influencing factors on human thermal comfort indoors.

2 Methods

2.1 MRT Delta Calculations for Windows Present in the IGDB Using the Spectrally-Resolved Method

The spectrally-resolved method was employed to calculate MRT delta values for the glazing products present in the IGDB using some default settings and assumptions. The calculation process and the settings for other variables are depicted in Fig. 1. For a more comprehensive understanding of the principles and calculation procedure, further information can be referred to our previous study [3].

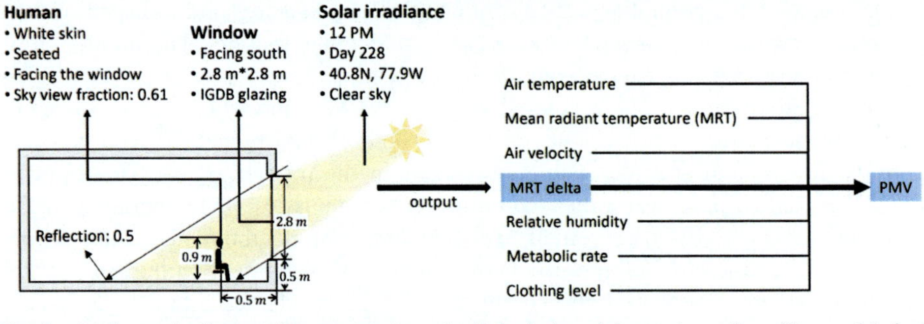

Fig. 1. A schematic diagram illustrating the solar light-induced thermal comfort effect and default settings in the calculations.

In the spectrally-resolved method, the solar irradiance, window transmittance, and skin absorptivity are combined by wavelength through superposition. To ensure consistency, the wavelength intervals of these three variables should be the same. To achieve this, interpolation and deletion of glazing transmittances were performed for certain wavelengths in the database. Specifically, wavelengths that were absent for more than 20 glazing products were excluded from the selected wavelength interval. Additionally, glazing with excessive missing wavelengths was also excluded to prevent excessive interpolation, which could potentially lead to inaccurate results. As a result, out of the total 5683 glazing products in the database, 5138 glazing products were selected for the calculations.

2.2 Calculations for the Preferred MRT Delta Values Under Various Indoor Environments

The original PMV is a seven-scale continuous index range from −3 to 3, which represents a scale from cold to hot [5]. People are generally regarded comfort when PMV values are in the range from −0.5 to 0.5 [5]. When other indoor environmental factors in the PMV model change, the preferred MRT delta value, which is the value of this additional factor needed to make people feel thermal comfort, would vary.

The preferred MRT delta values were calculated for specific indoor environmental settings listed in Table 1. It should be noted that air temperature and MRT were assumed the same in the analysis. The target PMV value for achieving the preferred MRT delta was set at 0. Along with the settings of the four original environmental variables in the PMV model, the metabolic rate of the human was set at 1 met and the clothing level was set at 0.6 clo during the calculation.

Table 1. Settings for the environmental variables.

Number	Air temperature (°C)	MRT (°C)	Air velocity (m/s)	Relative humidity (%)
E1	25	25	0.1	50
E2	22	22	0.1	50
E3	19	19	0.1	50
E4	22	22	0.2	50
E5	22	22	0	50
E6	22	22	0.1	30
E7	22	22	0.1	60

3 Results and Discussion

The MRT delta values of glazing products in the IGDB are shown in Fig. 2 in ascending order. The MRT delta values range from 0.0014 to 36.15, which corresponds to a difference of approximately 6.5 difference in PMV values. With the solar light-induced effect being considered and additional MRT delta value included in the model, the original PMV range is exceeded, which reflects the significant and influencing solar light-induced thermal effect of windows. In addition, the substantial variations in the glazing effects underscore the significant impact that windows have on the solar light-induced thermal comfort effect.

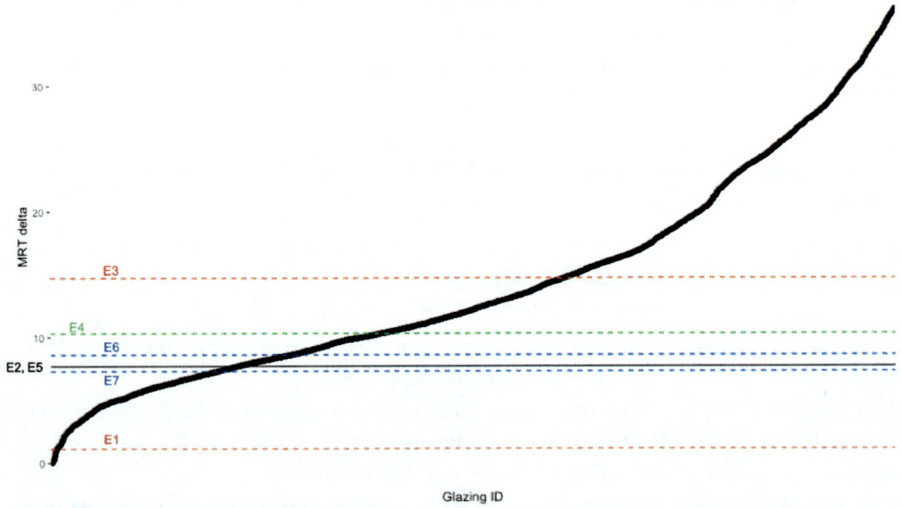

Number	E1	E2	E3	E4	E5	E6	E7
MRT delta	1.1	7.7	14.7	10.3	7.7	8.6	7.3

Fig. 2. The MRT delta values for glazing products present in the IGDB and the preferred MRT delta values for various indoor environments. The MRT delta values for the 5138 glazing products are arranged in ascending order. The ascending order of the MRT delta values, along with their large quantity, results in the accumulation of data points forming an ascending line (depicted in black). This line serves to visually represent distribution of MRT delta values for glazing products. The preferred MRT delta values for conditions E1 to E7, which are defined in Table 1, are indicated by horizontal dashed lines. Different colors are utilized to distinguish conditions with various settings: E2, serving as the baseline condition, is represented in black; E1 and E3 have different air temperature settings compared to E2, indicated in red; E4 and E5 have different air velocity settings, shown in green (since the MRT delta value of E5 is the same as E2, the line remains black); E6 and E7 have different Relative humidity settings, denoted by blue. The MRT values for these conditions are also summarized and listed at the bottom of the figure.

The preferred MRT delta values differ across various indoor environmental conditions, as shown in Fig. 2. E2 is the baseline condition with air temperature of 22°C, air velocity of 0.1 m/s, and relative humidity of 50%. With the increment of air temperature, the preferred MRT delta decreases. Conversely, an increase in air velocity leads to an increase in the preferred MRT delta. However, when the air velocity decreases from 0.1 m/s to 0 m/s, the preferred MRT delta remains unchanged. This indicates that low air velocity may counteract the additional solar effect. Regarding relative humidity, higher humidity results in a smaller preferred MRT delta, although the influence of relative humidity tends to be smaller compared to the other two variables.

The variations in the preferred MRT delta values across different indoor environments necessitate the selection of appropriate windows to achieve thermal comfort. Conversely, selecting different windows would require different indoor environment conditions, thereby offering the potential for energy savings when employing active strategies to maintain a comfortable indoor environment.

4 Conclusions

Glazing products available in the IGDB offer a diverse range of MRT delta values, reflecting the substantial role that windows have in solar light-induced thermal comfort effects. This effect can influence the requirements of other indoor environmental variables that are typically controlled by active strategies such as air conditioning systems in modern society. Therefore, a careful selection of window systems to use the solar light-induced thermal comfort effect in a more practical way will improve human thermal comfort indoors and enhance energy savings.

This study also presents a workflow and knowledge that can assist stakeholders and customers to evaluate and select window systems. The spectrally-resolved method, introduced in this paper and our previous publications, for calculating MRT delta values can be applied more broadly to evaluate target windows under specific solar, environmental, building, and human conditions. Furthermore, it would be beneficial if a thermal effect index for the windows in the IGDB could be developed and integrated in the National Fenestration Rating Council (NFRC) window rating system. This would offer customers direct information when choosing a glazing product in the market, as the thermal effect index would be displayed on the NFRC label. A higher index value would indicate a higher MRT delta value, implying that the glazing product allows more sunlight entering indoor spaces. Customers could then make informed decisions based on their specific needs and preferences. Through the generalization of the method and potentially incorporation of the index in the NFRC rating system, guidance related to the thermal effect of the windows could be provided to relavant organizations and customers. With that being said, ongoing research is still needed to gain a deeper understanding of the role of windows in the solar light-induced thermal comfort effects. Future research should explore more aspects to develop a more comprehensive and systematic rating system.

Acknowledgment. This project is supported by the NSF award: # 2001207: CAREER: Understanding the Thermal and Optical Behaviors of the Near Infrared (NIR)-Selective Dynamic Glazing Structures.

References

1. Zhao, Q., Lian, Z., Lai, D.: Thermal comfort models and their developments: a review. Energy Built Environ. **2**(1), 21–33 (2021)
2. Arens, E., Hoyt, T., Zhou, X., Huang, L., Zhang, H., Schiavon, S.: Modeling the comfort effects of short-wave solar radiation indoors. Build. Environ. **88**, 3–9 (2015)

3. Wang, N., Wang, J.: A spectrally-resolved method for evaluating the solar effect on user thermal comfort in the near-window zone. Build. Environ. **202**, 108044 (2021)
4. Berkeley Lab Homepage. https://windows.lbl.gov/software/igdb. Accessed 11 Apr 2023
5. Yau, Y., Chew, B.: A review on predicted mean vote and adaptive thermal comfort models. Build. Serv. Eng. Res. Technol. **35**(1), 23–35 (2014)

Nanoscale Photothermal Effects Driven by Solar Radiation for Building Envelope Thermal Behavior Management

Mohammad Elmi and Julian Wang(✉)

Department of Architectural Engineering, Pennsylvania State University, University Park, PA 16802, USA

jqw5965@psu.edu

Abstract. The passive solar design includes ways to utilize solar radiation and then maintain buildings' thermal comfort and help in reducing the energy consumption of buildings. Using and integrating photothermal materials into building envelopes or other components have been widely accepted in this domain. For instance, high solar absorptive materials added in sunspaces attached to the building façades in order to use solar heat and reduce energy consumption is one basic type of passive solar utilization since the 1970s. However, in recent years, some developments of new photothermal materials or structures, especially at the nanoscale, have exhibited unique solar-thermal conversions, which drove some new ways for solar utilization in buildings. In this work, we discussed photothermal effect and its applications in energy-efficient buildings in more detail and explored how nanoscale photothermal effects can help in thermal management in buildings.

Keywords: Photothermal effect · Nanoscale heat transfer · Energy-efficient buildings · Passive solar design · Building envelope · Building materials · SHGC

1 Introduction

Buildings and the construction industry made up about 36% of the total final energy use and about 39% of the energy-related CO2 emissions. Heating and cooling are responsible for approximately 77% of this energy consumption [1, 2]. In order to effectively reduce overall carbon emissions, building energy consumption must be reduced along with using cleaner energy sources [3, 4]. Passive building design is the primary and most significant part of energy-efficient construction. Passive solar heating and cooling are emerging as effective methods for reducing energy use and carbon emissions. The passive solar design incorporates techniques for preserving thermal comfort in buildings without the use of mechanical or electrical equipment [5–7]. Utilizing sunspaces in building envelopes is a basic passive solar design that has been used for more than 50 years.

D. Renné et al. (Eds.): ASES SOLAR 2023, SPE, pp. 27–32, 2023.
https://doi.org/10.1007/978-3-031-39147-7_4

Light-to-heat energy conversion or photothermal effect is a promising way to use solar energy in buildings and passive solar design. Recent developments in nano-micro materials can extremely help in this. There are a lot of research works dedicated to applications of nano-micro materials in biomedical science and nanoscale light-to-heat energy conversion utilization in biomedicine [8, 9]. But there is lack of research studies in photothermal, and specifically, nano photothermal effects applications in energy-efficient buildings.

Here we discuss photothermal effect and its potential applications in energy-efficient buildings with concentration on nanoscale photothermal effect.

2 Photothermal Effect

When sunlight strikes a material, it can be reflected, absorbed, or transmitted. Also, the absorbed photon can transfer its energy to material by different kinds of processes, which cause various energy releases. When absorbed photon's energy is transferred to an electron and is released as heat or kinetic energy, the photothermal effect occurs.

Photothermal effect itself has three different mechanisms, which can occur by one or more of them in each photothermal material: 1- plasmonic localized heating, 2- electron-hole generation and relaxation, and 3- thermal vibration of molecules [10]. Metallic materials like gold and silver are placed among the first group. Semiconductors' photothermal effect mostly occurs by electron-hole generation and relaxation, and other materials like carbon-based nanomaterials or polymers have the third mechanism, thermal vibration of molecules. We have discussed each photothermal mechanism in detail in our previous work [11].

2.1 Nanoscale Photothermal Effect

When the frequency of the incoming light matches with the natural frequency of the electrons on the surface of the metal, localized surface plasmon resonance takes place. This excites the electrons, and hot electrons are produced as a result. As a result of the collective oscillation of the excited hot electrons and the incoming electromagnetic field, heat is produced. By electron-electron scattering, hot electrons redistribute their energy and heat is generated in the plasmonic element [10–12]. Figure 1 depicts the schematic of this process.

Mostly, these plasmonic elements are metallic nanoparticles. Numerous studies on plasmonic nanoparticles and their photothermal applications have been conducted recently [13, 14]. Plasmonic nanoparticles like gold and silver nanoparticles have high photothermal efficiency because of their high absorption capability.

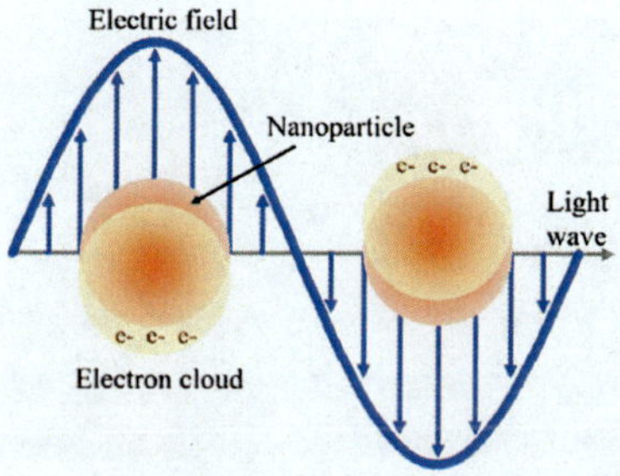

Fig. 1. Plasmonic localized heating process schematic

3 Photothermal Effect Applications in Energy-Efficient Buildings

Based on our previous work [11], generally we can categorize photothermal effect applications in buildings to three main groups: 1- Photothermal materials applied to improve window's thermal performance, 2- Photothermal effect in photo-thermochromic and phase change materials used in buildings envelopes, and 3- Solar-thermal conversion in conventional passive solar designs.

Nanoparticles can be useful for the first and the second group of applications. But plasmonic nanoparticles are mostly suitable for the first one, photothermal materials applied to improve window's thermal performance. For this purpose, plasmonic nanoparticles like Au nanospheres, should be deposited on the window's surface. A glass sample coated by 60 nm gold nanospheres and a bare glass are shown in Fig. 2. When sunlight strikes the coated window, photothermal effect occurs by the deposited nanoparticles and the surface temperature of the window increases. Then, the generated heat is transferred to the inside by radiation heat transfer. To show this in more detail and experimentally, two samples, one with Au nanospheres coating and one without any coating, were prepared and were exposed to the light source. The total irradiance on the samples' surfaces was 80 W/m^2, which is about 1/13 of the solar total irradiance on the earth's surface. Samples were held under light exposure for 3 h. Figure 3 shows the thermal image of the samples after 3 h. As is depicted in Fig. 3, the temperature rise of the coated glass is more than the bare glass. There is 2 °C temperature difference between samples in average. Although, in comparison with the solar light, the total irradiance on the samples' surfaces was very low, photothermal effect by Au nanospheres could make a sensible temperature difference between coated and uncoated samples.

The Solar Heat Gain Coefficient (SHGC) of a window is calculated by:

$$SHGC = \tau_s + N_i \alpha_s \quad (1)$$

where N_i is the inward-flowing fraction of the absorbed radiation, τ_s and α_s are the solar transmittance and absorptance of the window, respectively [15]. N_i shows how much of the absorbed energy is radiated inside, and how much of it is radiated outside of the room.

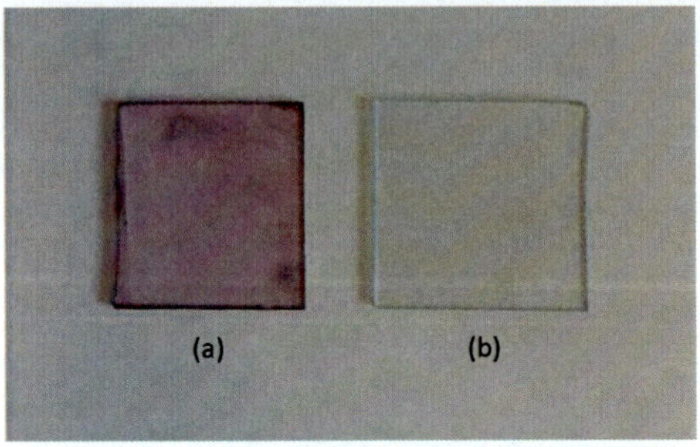

Fig. 2. (a): Gold nanoparticle coated glass, (b): Bare glass

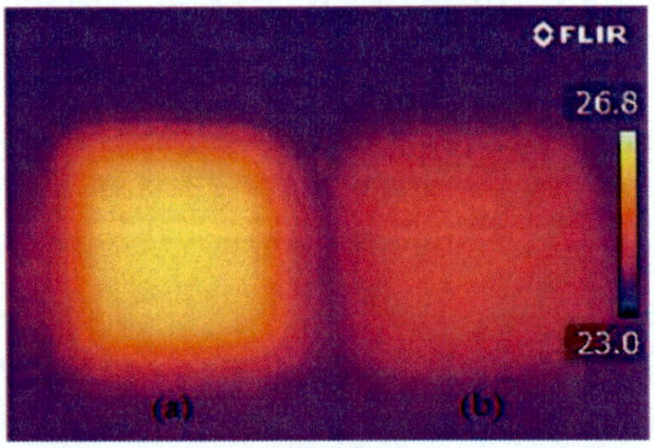

Fig. 3. Temperature map of the samples after the light exposure (a): Gold nanoparticle coated glass, (b): Bare glass.

Previous studies show that utilizing plasmonic nanoparticles on windows can increase the SHGC of windows in comparison with Low-E windows [15]. For example, Zhang et al. theoretically showed that solar heat gain through windows in photothermal material coated windows is about 15% more than Low-E windows and can reduce heat loss about 16% more. Also, they showed that using photothermal material coated windows can reduce heat loss by about 38% relative to uncoated windows [15]. The transmittance and absorptance spectrum of the Au nanoparticle coated glass sample and a bare glass are demonstrated in Fig. 4. As is shown in Fig. 4, the absorption of the glass is extremely increased by Au nanoparticle deposition, and the absorptance spectrum has a peak on the wavelength of 550 nm.

We believe the increase in the SHGC is not only because of the change in optical spectra of the window's surface, but also photothermal effect by coated plasmonic nanoparticles changes the inward-flowing fraction of the absorbed radiation, N_i, and it can help us in controlling the heat transfer in building envelopes. But still more experimental research works are required to investigate that.

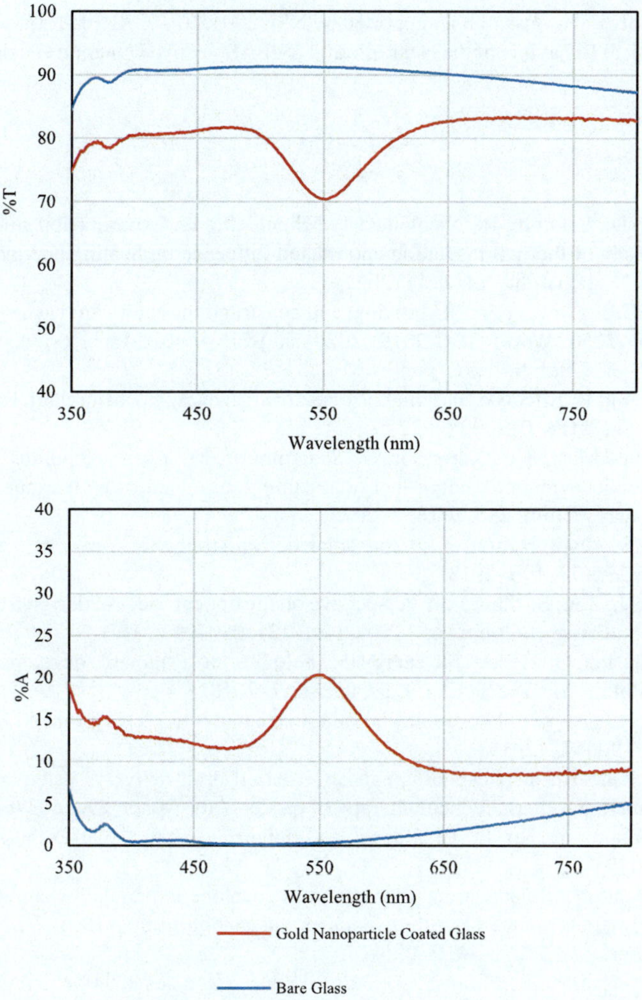

Fig. 4. Optical spectra of the gold nanoparticle coated glass and a bare glass

4 Conclusion and Future Research Works

Here we showed how photothermal effects by plasmonic nanoparticles can enhance temperature rise in coated windows when they are exposed to the sunlight, which will result in thermal management and energy saving in buildings. By recent advances in photothermal materials and plasmonic nanoparticles, light-to-heat energy conversion is exhibiting enormous potential of building energy savings. Coating plasmonic nanoparticles on windows, besides changing the optical properties of the windows, can lead to some changes in their thermal properties as well. Studying the effect of plasmonic nanoparticles on the windows thermal properties is necessary in the field, which requires more theoretical and experimental research works.

Acknowledgments. This research is supported by NSF 2001207 – CAREER: Understanding the Thermal and Optical Behaviors of the Near Infrared (NIR)-Selective Dynamic Glazing Structures.

References

1. D'Agostino, D., Landolfi, R., Nicolella, M., Minichiello, F.: Experimental study on the performance decay of thermal insulation and related influence on heating energy consumption in buildings. Sustainability **14**, 2947 (2022)
2. Council, W.G.B.: New report: the building and construction sector can reach net zero carbon emissions by 2050. World GBC 2019 (2020). https://www.worldgbc.org/news-media/WorldGBC-embodied-carbon-report-published
3. Yang, Y., Wang, P.: Effects of building physics form on energy consumption for buildings. J. Phys. Conf. Ser. **2186**, 012008 (2022)
4. Laskari, M., de Masi, R.F., Karatasou, S., Santamouris, M., Assimakopoulos, M.N.: On the impact of user behaviour on heating energy consumption and indoor temperature in residential buildings. Energy Build. **255**, 111657 (2022)
5. Stevanović, S.: Optimization of passive solar design strategies: a review. Renew. Sustain. Energy Rev. **25**, 177–196 (2013)
6. Jiang, Y., Li, J., Zhu, B., Zhu, J.: A Perspective on the optical spectral design for passive solar heating and radiative cooling. Appl. Phys. Lett. **121**, 090501 (2022)
7. Chel, A., Kaushik, G.: Renewable energy technologies for sustainable development of energy efficient building. Alexandria Eng. J. **57**, 655–669 (2018)
8. Kim, M., Lee, J., Nam, J.: Plasmonic photothermal nanoparticles for biomedical applications. Adv. Sci. **6**, 1900471 (2019)
9. Chen, W., et al.: Black phosphorus nanosheet-based drug delivery system for synergistic photodynamic/photothermal/chemotherapy of cancer. Adv. Mater. **29**, 1603864 (2017)
10. Chen, C., Kuang, Y., Hu, L.: Challenges and opportunities for solar evaporation. Joule **3**, 683–718 (2019)
11. Elmi, M., Wang, J.: Solar-thermal conversion in envelope materials for energy savings. In: Advanced Materials in Smart Building Skins for Sustainability, pp. 113–127. Springer (2023). https://doi.org/10.1007/978-3-031-09695-2_5
12. Anis, S.F., Hashaikeh, R., Hilal, N.: Functional materials in desalination: a review. Desalination **468**, 114077 (2019)
13. Ai, K., et al.: Localized surface plasmon resonance properties and biomedical applications of copper selenide nanomaterials. Mater. Today Chem. **20**, 100402 (2021)
14. Wang, J., Shi, D.: Spectral selective and photothermal nano structured thin films for energy efficient windows. Appl. Energy **208**, 83–96 (2017)
15. Zhang, E., Duan, Q., Wang, J., Zhao, Y., Feng, Y.: Experimental and numerical analysis of the energy performance of building windows with solar NIR-driven plasmonic photothermal effects. Energy Convers. Manag. **245**, 114594 (2021)

Vernacular Approach to Passive Solar Design

Sandhya Sood(✉)

Principal Architect, Accent Architecture+Design, Berkeley, USA
info@accentarchitecture.com

Abstract. "Vernacular" (adj.) derives from the Latin word "vernaculus", meaning, "native to a country, domestic, indigenous" [1]. When pertaining to building, it refers to architecture that belongs to a place and meets the needs of its ordinary people and their culture. Vernacular architecture responds to the resources of its land, such as climate and geographic features, by using traditional materials and techniques for construction [2]. Many vernacular building typologies, including dwellings are inherently climate adaptive and utilize passive solar energy by taking advantage of their built form, material properties and solar orientation, among other attributes. They harness the sun's passive energy for natural illumination, ventilation, and to keep living spaces warmer or cooler than outside. These sustainable design interventions are simple and affordable with negligible installation costs as they are integral to the building design. In 2021, the U.S residential sector consumed 20.9 quadrillion Btu of primary energy, almost 22% of U.S primary energy consumption [3]. According to the 2019 American Housing Survey, almost 90% of all buildings in the US are single family houses [4]. Clearly, passive solar design continues to be relevant in residential architecture as it can be applied to conserve active energy use or where energy access is a challenge. It would be beneficial to identify strategies from different parts of the world for our understanding, their preservation and implementation in new buildings or existing ones. This paper summarizes the working, climatic response, design and advantages of three passive solar design strategies found in vernacular architecture. It also demonstrates how the strategies have been successfully incorporated in historic and contemporary houses designed by architects, including, the author.

Keywords: Passive solar design · Vernacular architecture · Energy conserving strategies

1 Thermal Battery to Store and Distribute Passive Solar Energy

1.1 Vernacular Architecture of Dwellings in the Himalayas, India

Vernacular dwellings in the rocky terrain of the Himalayas are of several types, their built tradition often passed within resident families and their local community. Built mostly with stone, wood and slate, the variation in masonry and roof pitch depends on the altitude and local climate. A well-recognized indigenous construction technique found in Himachal Pradesh, called Kath-Khuni, entails dry stacking flat stones alternately with

D. Renné et al. (Eds.): ASES SOLAR 2023, SPE, pp. 33–42, 2023.
https://doi.org/10.1007/978-3-031-39147-7_5

long sections of wood in a double skin wall with loose fill [5]. These walls form the base of the structure with lighter upper floors built primarily in wood, often carved or decorated to reflect folk culture (Fig. 1).The house interior is sometimes plastered with mud, adding to the limited vocabulary of materials contrasting in texture, color and densities. Thermal mass of double skinned walls helps retain heat in a mountainous region that experiences freezing temperatures and a harsh climate in winter.

Fig. 1. Kath-Khuni wall construction at the lower floor of a house
Photo: Sahapedia

1.2 Contemporary Example in the Foothills of the Himalayas, North India

The strategy of using the building as a thermal battery, found in vernacular architecture, is incorporated in the design of a zero-waste, low energy residential retreat located in the Himalayan foothills. Daytime temperature in summer ranges from 80F to 50F while winter lows can be close to freezing. Construction was done by hand, in the late 1990's, using small tools. This method enabled conservation of the site's topography and ecology (Fig. 2). Undressed sandstone was bought from a neighboring quarry while large stone slabs were sourced from a local village. 18″ deep rubble masonry cavity walls laid with cement-sand mortar are built between reinforced concrete columns faced with stone. The cavity is packed with rammed earth and stone debris, as part of an integrated construction technique (Fig. 3). The insulated roof slab is cast with reinforced concrete, a modern material that adds wind resistance to a native material vocabulary. The thermal mass of slate clad roof, stone clad floor and double skinned stone walls acts as a natural heat sink, reducing diurnal temperature differences indoors. Openings are recessed and highlighted with vernacular Himalayan motifs, assimilating alpine cultural expression with the new construction (Fig. 4).

Fig. 2. Building with small tools
Photo: Sandhya Sood

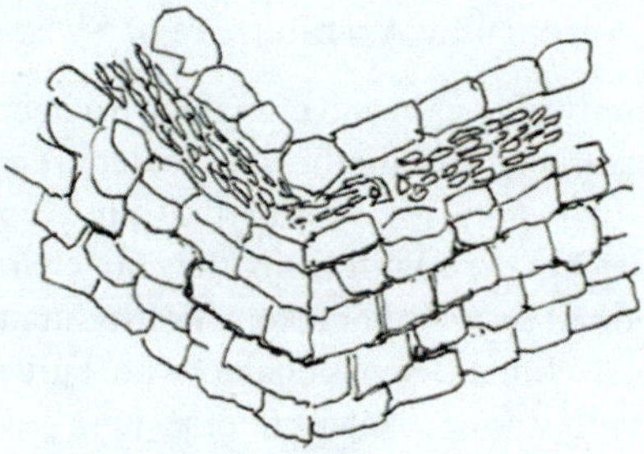

Fig. 3. Random rubble cavity wall
Drawing: Sandhya Sood

Fig. 4. House in the Himalayan foothills designed by architect Sandhya Sood
Photo: Sandhya Sood

How the Strategy Works

The strategy of using the building's mass as thermal battery employs the principle of time lag. During daytime hours, the building stores heat from the sun in materials that form the walls, floor slab and roof slab. In the evening, when it is cooler outside, the stored heat is transferred to adjoining interior spaces, making inhabitants feel warmer. The thermal mass absorbs and holds heat due to the high thermal density of its materials, delaying heat transfer from outside to inside. The house adapts to warm summer days with a shaded verandah and natural ventilation through operable windows (Fig. 4). Openings are recessed in the 18″ thick stone walls, blocking high summer sun but permitting it in winter, when it is lower in the sky.

2 Natural Cooling by Convection and Ventilation

2.1 Vernacular Architecture of Swiss Chalets in Switzerland

Diversity of terrain and climate in Switzerland influenced the pitch of roofs and selection of materials forming the vernacular architecture of early Swiss chalets. These emerged over two centuries ago as rural huts, cottages or small cabins [6]. Usually square or compact in their built form, they are characterized by wide gable roofs with broad eaves supported by wood brackets. Protruding balconies or galleries protect lower levels from adverse climate. Swiss chalets were built with timber or interlocking logs and occassionally with stone for lower floors where pasture-fed cattle were herded [6], typically in the summer (Fig. 5). Operable windows enable natural ventilation during warm summers (Fig. 6). The tradition of building flower boxes at windows, elaborate brackets or motifs at railings emphasizes a picturesque landscape.

Fig. 5. Rustic chalet in Nendaz, Switzerland
Photo: Sandhya Sood

Fig. 6. Swiss chalet in Nendaz, Switzerland
Photo: Sandhya Sood

2.2 Historic Example in Berkeley, CA

Julia Morgan, known for her vernacular approach to California's regional architecture [7] designed a chalet-inspired house in Berkeley, in 1910, for Mr. Sharp [8]. Chalet architecture is often expressed through profuse windows and balconies. The genesis of this compact, two-story house, is an interior balcony at the center of a square plan. Oriented west, it captures golden daylight from sunsets over the Pacific Ocean (Fig. 7). Berkeley summers are rather dry, with daytime temperature typically ranging from 52F to 72F, but soaring upwards of 85F in recent years. When local climate shifts to hotter days, passive cooling improves the thermal comfort of this house with no power consumption, using natural ventilation as an effective strategy (Fig. 8).

Fig. 7. Living room at sunset
Photo: Sandhya Sood

Fig. 8. Operable windows under roof
Photo: Sandhya Sood

How the Strategy Works
Heat from the sun penetrates the exterior wood framed walls, making the house warm. The strategy of passive cooling relies on warm air that rises by convection and exits through an opening at the roof, thereby cooling the spaces below (Fig. 8). Cooler outside air pulls through windows at the lower level, replacing rising warm air with fresh, cool air (Fig. 9). On hot days, the upper level is warmer due to temperature stratification and air is allowed to escape by stack ventilation [9]. The balcony provides an opportunity to cool an inhabitant standing in the path of cross-flow ventilation. This occurs when the casement windows above the balcony are open (Fig. 8). The intuitive knowledge of operating windows is intergenerational as it was passed on to the current owner by his parents when he was a child, growing up in the house.

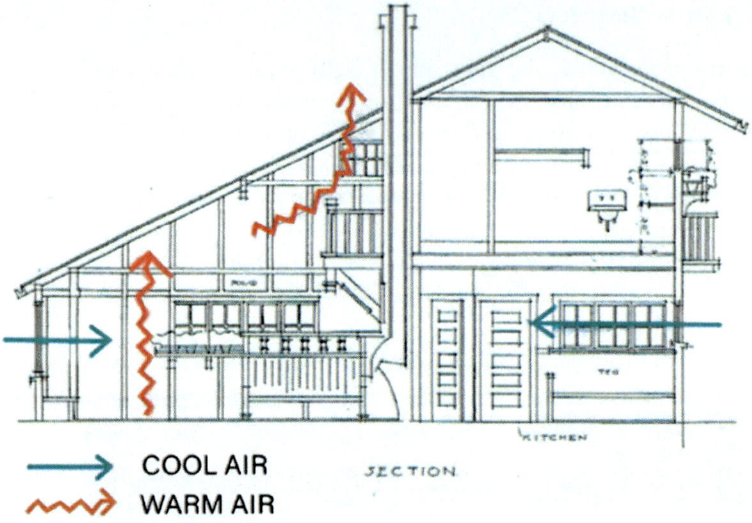

Fig. 9. Cross section drawing of house designed by architect Julia Morgan Diagram: Sandhya Sood

3 Solar Orientation for Natural Light

3.1 Vernacular Architecture of Ranch Houses in California

An example of vernacular architecture that takes advantage of solar orientation is California's quintessential Ranch House. Ranch houses evolved into a suburban postwar typology to support an informal culture of life under the sun, emblematic of California's pastoral landscape [10]. Several features such as post and beam construction, bedrooms in wings on the sides and family spaces in the center, leading to a courtyard or garden, are identifiable. This spatial arrangement resulted in a building typically one room deep, with access to sunlight from most orientations (Fig. 10). Finish materials range from wood shakes to clay roof tiles, stucco to board and batten; juxtaposed with courtyards and patios. Several variations in floor layouts can be found, from rooms sprawled across acres of land to smaller structures with deep overhangs. Cliff May is considered a pioneer in the design of regional iterations of California's ranch houses [10]. Not all ranch houses embody the original spirit and many, such as the example below, have the potential to imbibe solar benefits of their predecessors.

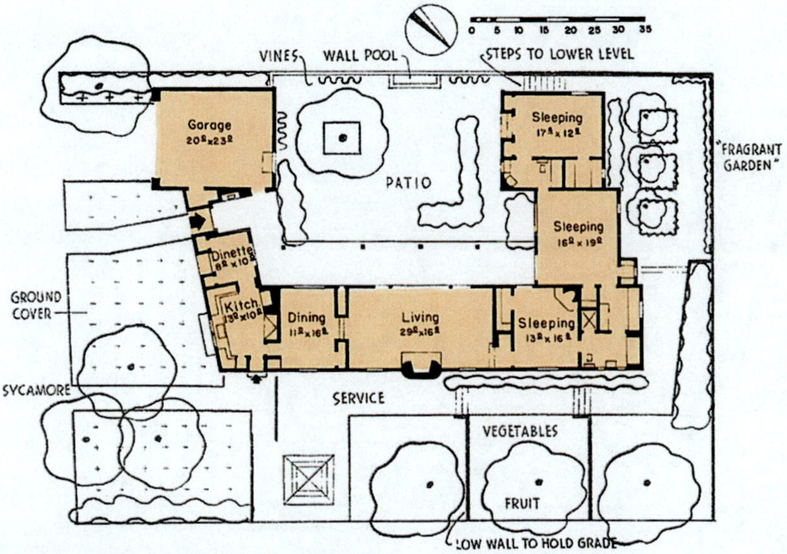

Fig. 10. Ranch house plan by Cliff May
Courtesy: Sunset Publishing Corp.

Fig. 11. Mid-century ranch house in Los Altos, CA
Photo: Sandhya Sood

3.2 Contemporary Example in Los Altos, CA

A mid-century ranch house (Fig. 11) renovated over the decades had lost the relationship to it's south garden. An addition to the house in 2009 became an opportunity to peel away incongrous windows that did not bring in adequate daylight. The south facing addition was designed to wrap around the existing south wing, taking advantage of its solar orientation (Fig. 12). Daylight permeates through strategically placed and sized windows and openings, nestled within this "wrapped" wing (Fig. 13). The addition reinforces the ranch house typology, connecting the house to it's site and landscape.

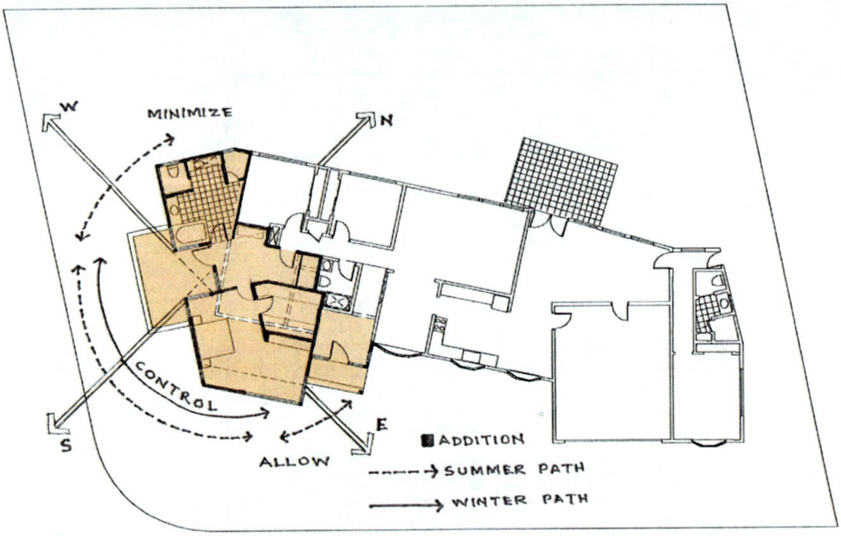

Fig. 12. Sun paths at south wing of mid-century ranch house in Los Altos, CA. Drawing: Sandhya Sood.

How the Strategy Works

The diagram shows movement of the sun around the new south wing in summer and winter (Fig. 12). Openings in the center of this new wing are partially or fully shaded from the afternoon sun in summer as they are setback from the adjoining spatial volumes, bringing glare free light to the interior (Fig. 13). Sunlight is modulated by the shape, size and placement of windows in the new bedroom and bathroom that are also equipped with operable window shades. Bedroom windows facing south are tall and narrow whereas those facing east are larger to permit early morning light, welcome by inhabitants. In winter, the family gathers in the south patio to soak in the sun, its warmth extending to the adjoining garden (Fig. 14). Houses with clerestory, a range of window sizes and shading devices can allow or prevent sunlight to penetrate appropriately arranged, habitable spaces. Quality and intensity of natural light can vary from sharp to diffused, depending on solar orientation as well as microclimate, latitude, longitude and other factors.

Fig. 13. Clerestory window at south addition of mid-century ranch house in Los Altos, CA. Photo: Rachel Schuler

Fig. 14. South-southwest facing openings are setback, shaded by adjacent walls and trellis, at center of the addition designed by architect Sandhya Sood in Los Altos, CA.
Photo: Rachel Schuler

4 Conclusion

People that inhabit vernacular architecture engage with its embodied bioclimatic response in ways that are meaningful and useful to their everyday lives or culture. When passive solar design is integrated with attributes of vernacular architecture, it can provide added benefits of enhanced habitability and connection to place.

Passive solar design supports sustainability as it can be a significant ally in mitigating the rapid rate of increase of carbon emissions. It can contribute positively to diminishing the pressure on global energy demand as well as reduce costs. Although supplanting natural light with artificial illumination may not yield the same visual, thermal or psychological benefits, passive solar design strategies can be implemented in conjunction with electrical lighting. They can also be used with other energy efficiency solutions such as insulation and thermal barriers or be supplemented with mechanical heating or cooling. Unless all stakeholders understand the value of harnessing the sun's passive energy as a critical counterpart to energy generation and renewal, our path forward to an equitable energy future will be challenging.

References

1. Douglas, H.: Etymology of vernacular, Online Etymology Dictionary. https://www.etymonline.com/word/vernacular. Accessed 29 May 2023
2. Rapoport, A.: House Form and Culture, 1st.edn. Prentice-Hall, Englewood Cliffs (1969)
3. CSS, University of Michigan. https://css.umich.edu/publications/factsheets/energy/us-energy-system-factsheet. Accessed 06 June 2023
4. Potter, B.: Construction Physics, Analysis of the US Building stock (2019 American Housing Survey), Construction physics. https://www.construction-physics.com/p/every-building-in-america-an-analysis. Accessed 17 May 2023
5. Shah M., Thakkar J.: The Himalayan Vernacular: Kath Khuni Architecture. Sahapedia (2018). https://www.sahapedia.org/the-himalayan-vernacular-kath-khuni-architecture/. Accessed 17 May 2023
6. Giberti, B.: The chalet as archetype: the bungalow, picturesque tradition and vernacular form. TDSR J. **3**(I), 57 (1991)
7. Sood S.: 56 years after her death, Julia Morgan wins top award, Berkeleyside. http://www.berkeleyside.com/2014/01/17/56-years-after-her-death-julia-morgan-wins-top-award/. Accessed 17 May 2023
8. Sood S.: Julia Morgan: Architecture for Sustainability. Landmarks California (2012). http://www.accentarchitecture.com/. Accessed 04 06/04/23
9. Leo S., Shiva N., Maiya. M.P.: Passive alternatives to mechanical air conditioning of building: a review. Build. Environ. **66**, 54–64 (2013)
10. Sunset Magazine, May, C.: Sunset Western Ranch Houses, 1st edn. Lane Publishing Co. (1946)

Assessment of the Window Performance from the Light Provision and Circadian Light Aspects

Neda Ghaeili[1](✉), Shevvaa Beiglary[1], Julian Wang[1], and Ying-Ling Jao[2]

[1] Department of Architectural Engineering, Pennsylvania State University, State College, USA
nxg51@psu.edu

[2] Ross and Carol Nese College of Nursing, Pennsylvania State University, University Park, PA, USA

Abstract. Various mechanisms and structures have been developed to enhance the energy performance of architectural windows, taking into account the connection between light and circadian rhythm. This correlation places additional demands on window energy performance. Regulating the circadian cycle depends on exposure to specific wavelengths of light at particular times. Electric lights are often used to achieve and maintain the desired level of circadian light stimulus, which increases the energy load on lighting systems. However, if windows allow the appropriate solar spectra into a room, the broad daylight spectrum could provide sufficient indoor circadian light. The visible transmittance of glazing is vital for circadian light inlets. However, the incoming light may meet visible light requirements but not circadian light requirements. In such cases, electric light is used as a supplement, leading to increased energy consumption. This study aims to investigate the circadian light performance of various glazing systems using the International Glazing Database and simulations through the LARK plugin for Grasshopper. The selected glazing systems' photopic and circadian transmittance characteristics are analyzed and compared concerning indoor health's circadian light requirements, considering the use of indoor electric lights. Preliminary evidence from this study suggests that well-designed architectural window systems can provide sufficient circadian lighting conditions by transmitting solar and sky radiation at appropriate quantities and wavelengths. However, optimization challenges exist when balancing both photopic light provision and circadian light performance.

Keywords: Daylight · Architectural windows · Circadian light · Indoor health · Spectral analysis

1 Introduction

In recent years, the building envelope system has gained a great deal of attention due to its significant role in maintaining a comfortable indoor thermal environment and energy consumption by buildings. Due to its high heat transfer coefficient, the window

D. Renné et al. (Eds.): ASES SOLAR 2023, SPE, pp. 43–51, 2023.
https://doi.org/10.1007/978-3-031-39147-7_6

and glazing system represents the weakest thermal component of the building envelope and contributes approximately 60% of a building's overall energy consumption (Sun et al. 2018). A reliable and widely used energy performance rating system for building windows in North America has been established by the National Fenestration Rating Council (NFRC). Regarding energy efficiency and indoor environmental quality, the U-factor, Solar Heat Gain Coefficient (SHGC), and Visible Transmittance (Tvis) are the three most significant window performance properties (Feng, Duan, Wang, & Baur, 2020). A great deal of research has been conducted on the properties of windows and the energy performance of buildings, and these properties are now being used as indicators for selecting the right window system in light of different climates or utility demands of buildings. For example, it has been found (Chaiyapinunt and Khamporn 2009) that the transmittance of single-pane windows correlates with the predicted percentage of dissatisfaction (PPD) caused by solar radiation ($PPD_{solar} = 98.853T - 1.0723$). Alternatively, as discussed as (He, Thomas Ng, Hossain, & Skitmore, 2019)for cold climates, the window system with the low U-factor and high SHGC is more energy-efficient for building.

Reporting scientific research findings; there have been recent discoveries related to ipRGCs in the retina, which displays melanopsin and is responsible for providing the non-image-forming visual signals to the brain (Yamakawa, Tsujimura, & Okajima, 2019) and, by receiving light signals, triggers a psychological or physiological response in the body. Based on light exposure duration, specific physical and psychological reactions occur (Houser and Esposito 2021). The duration should be based on the 24-h cycle of sunrise and sunset to maintain a healthy psychophysical response called circadian health. The cycle diversion is associated with several health problems, including cancer, diabetes, cardiovascular disease, and insomnia. (Potočnik and Košir, 2021). Research conducted in sociology has established that the vast majority of Americans, approximately 87%, tend to spend the majority of their time indoors (Dodge et al. 2012). Likely, this percentage has significantly risen during the pandemic. Indoor lighting is one of the parameters for providing healthy indoors and categorizes into two categories: artificial light and daylighting. While, due to the controllability of the artificial light, it has been subjected to various studies regarding on its impact on the circadian health (Esaki et al. 2023), (Jao et al. 2022), (Zhang et al. 2023); the daylighting impacts on the circadian health has been subjected for investigation fewer than the artificial light. In the studies subjected to the circadian daylight, parameters such as the window's properties such as the visible transmittance, window color, number glazing in the window system, window direction, and window-to-wall ratio were evaluated to distinguish their impact on the inlet circadian light. So far, according to (Xiao et al. 2021), high Tvis does not guarantee high inlet circadian light; however, it is required for windows. Furthermore, regarding the glazing color, as long the glazing color does not impact the Tvis, all tinted windows inlet minimum level of needed circadian light. However, blue-tinted glasses inlet more circadian-weighted light and the glasses with red and orange colors do in fewer portions (Hartman 2022), (Chen 2019). Besides, the south-facing windows with a bigger window-to-wall ratio provide more circadian light; about the WWR; however, as reported by (Acosta 2017) (Acosta 2019) after the specific percentage of WWR according to the room condition, there is no significant difference between the different WWR and inlet circadian light; and all WWRs inlet same amount of circadian light.

In all conducted studies regarding window performance in transmitting circadian light, windows' failure to provide the circadian light and supplement the artificial light to fulfill this failure is missing. Therefore, in this study, the performance of thirteen windows from the energy and inlet circadian light are investigated to distinguish the relation between them.

2 Methodology

This work analyzed thirteen windows regarding their capabilities for meeting circadian health standards. The International Glazing Data Base (IGDB) was used to select thirteen glazings based on their SPD, and the intention was to select the windows with different peaks of visible transmittance. Figure 1 and Table 1 illustrate these windows' spectral transmittance curves and critical properties respectively. As listed in Table 1, the selected windows are in a pair regarding their Tvis with different Tc to assess this optical indicator's usability in the evaluation of circadian light.

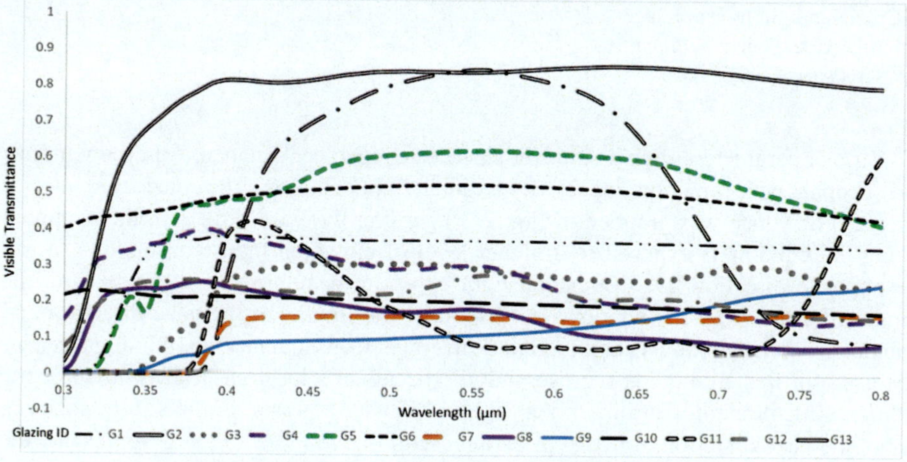

Fig. 1. Spectral transmittance of the selected windows

Lark Plugin for Grasshopper was used for analyzing the circadian window performance. A 2.8 * 2.8 * 2.8 m^3 box model was used for this work. Only the south-facing façade contains a window with a 30% window-to-wall ratio. The simulation was conducted in ASHRAE climate zones 4 in Denver, CO.

Table 1. Properties of chosen windows

ID	Color appearance	Tvis*	Tc**	SHGC***	U-factor	Peak Tvis
G1	Silver	0.361	0.37	0.476	5.538	400
G2	Clear	0.361	0.75	0.68	5.813	475
G3	Grey	0.272	0.28	0.281	3.222	510
G4	NA****	0.272	0.35	0.38	5.698	470
G5	Green	0.514	0.6	0.475	5.809	560
G6	Silver Grey	0.514	0.52	0.58	3.454	555
G7	Bronze	0.159	0.15	0.374	5.757	750
G8	NA	0.159	0.22	0.348	5.404	470
G9	Dark blue	0.2	0.12	0.503	5.834	780
G10	Silver	0.2	0.21	0.334	5.685	395
G11	Clear	0.234	0.11	0.703	5.818	780
G12	Grey	0.234	0.24	0.227	3.178	560
G13	Clear Low_E	0.844	0.83	0.794	5.821	630

* Visible transmittance,
** Circadian light transmittance,
*** Solar Heat Gain Coefficient,
*** * not specified

Various parameters such as interior surface reflectance, distance from the window, and occupant gaze direction impact the circadian light exposure. The considered interior surface reflectance was graphed in Fig. 2. Regarding the gaze direction, the occupant facing the window is exposed to a higher level of circadian light than the one facing the opposite direction. Less circadian light exposure occurs in the distance from the window, farther from the window. In this study, the impact of the gaze direction and distance from the window is assumed to be average. Consequently, an individual sensor was used for this simulation representing the occupant's location in the middle of the room at 120cm height from the floor_ at eye level for a person in the sitting position. About the gaze direction, it had kept parallel with the window. Figure 3 represents the plan of the considered model with the sensor's location and gaze direction.

As reported in WELL standards (Circadian Lighting Design, 2022) and (Yun et al. 2021) for an office, a well-lit room should achieve above 250 lx m-EDI between 9 a.m. to 1 p.m. Therefore, for this simulation, the same time window in the entire year was considered for the simulation, and the threshold for evaluating the window performance was the m-EDI above 250 lx.

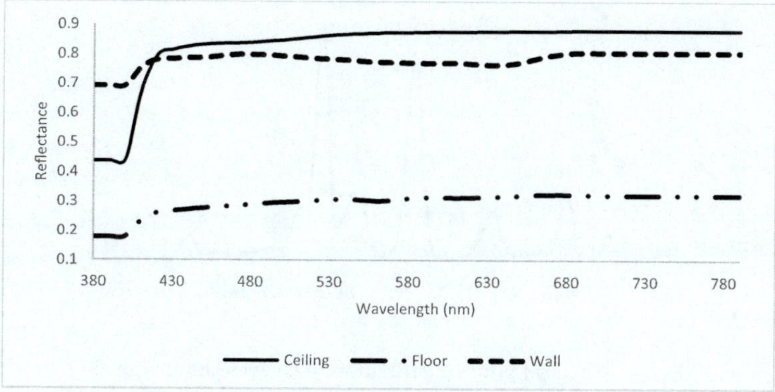

Fig. 2. Spectral characteristic of the material used for floor, ceiling, and interior walls

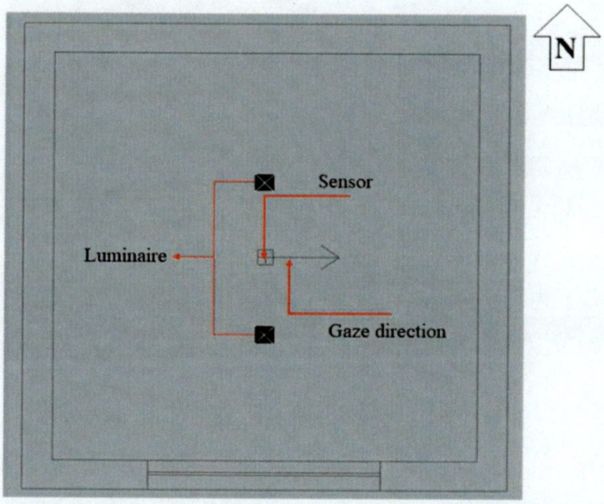

Fig. 3. Plan view of the sensor location and direction

3 Result

As discussed in the Method, there is a threshold of 250 lx for m-EDI and the requirement that the space achieves this level of light from the morning until noon. In this work, the SPD of selected windows was exported from the Optics program and used in the Lark to simulate the window performance. Figure 5 shows the circadian performance of these windows.

In total, the windows' performance was simulated between 9 a.m. to 1 p.m. with 6 min intervals that made 14600 timesteps. As listed in Fig. 5, windows G2, G4,G5, G6, G12, and G13 with m-EDI over the 250lx respectively for 97.87%, 89.14%, 93.47%, 94.73%, 77.09%, and 93.21%, of entire year performance desirably. While G7 and G9, by 9.88% and 11.21% m-EDI above the threshold during the year, fail to meet the requirement.

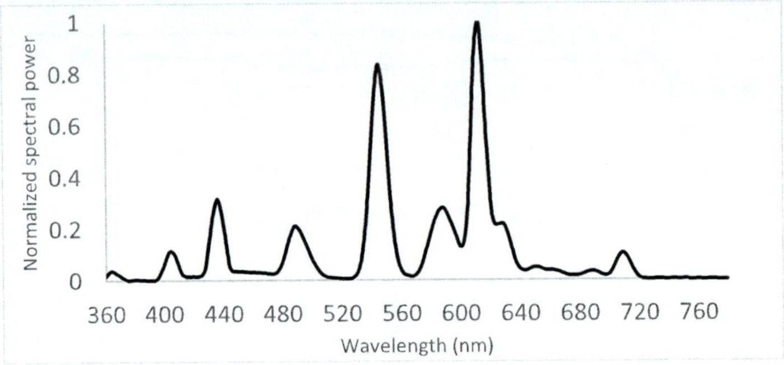

Fig. 4. Spectral power distribution of selected luminaire

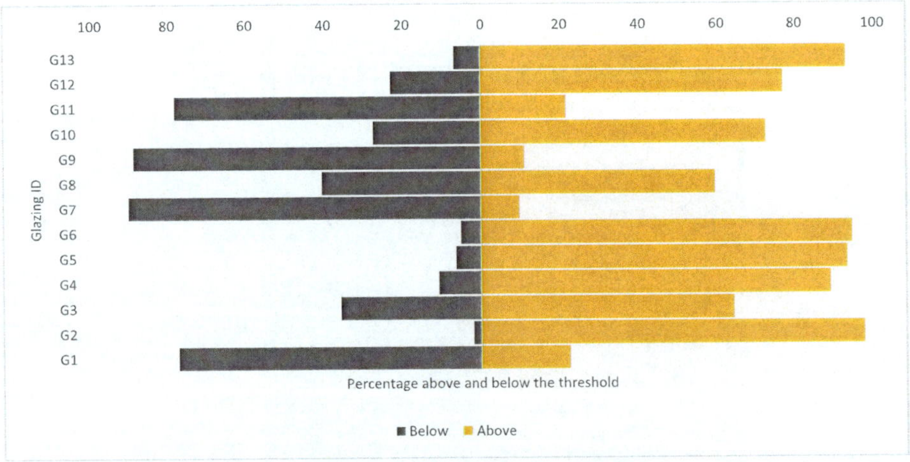

Fig. 5. Performance of Windows in relation to circadian daylighting

Consequently, using the electric light is demanded to fulfill the portion of the year when the window does not meet the threshold.

In this work, luminaires of TBS600 1x49W D7 TL5/840 HFP at the two points of the ceiling were considered for the artificial light. The locations of the two points in plan and spectral power distribution of the luminaire are presented in Figs. 3 and 4, respectively. Keeping lights on improves the room's circadian condition, and m-EDI listed more timesteps above 250 lx. Figure 6 represents the impact of electric light on indoor circadian light.

For models with Windows G2, G4, G5, G6, G10, G12, and G13, using the luminaire does not impact the circadian performance of the model so much. For the model with windows G2, G4, G5, G6, G12, and G13, it is evident that the solo performance of the window is sufficient to meet the requirements, and there is no need for extra artificial lighting. However, the model with window G10 which uses the electric light increases the timesteps above the threshold from 10597 to 10686, shows the type and SPD of the chosen luminaire is also essential factor for providing a healthy circadian environment.

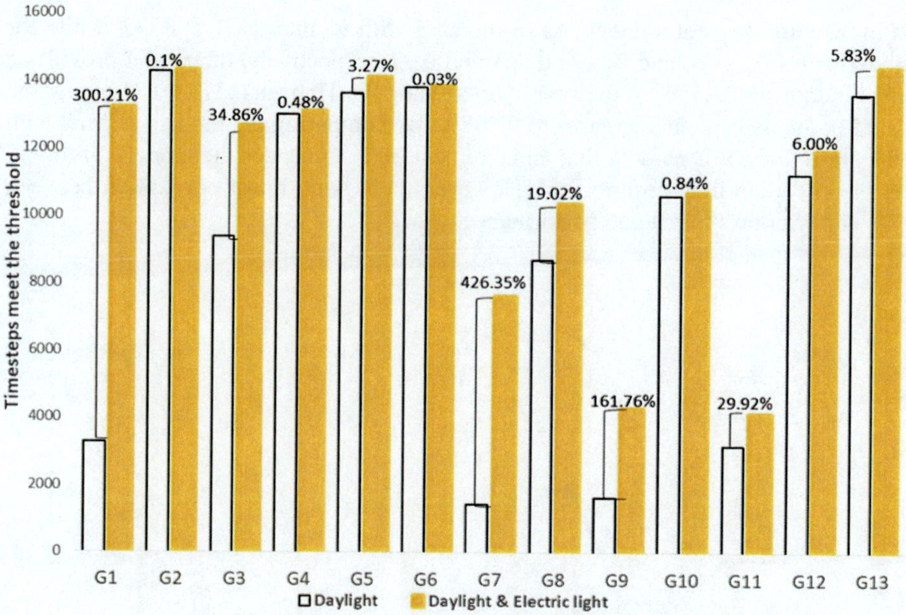

Fig. 6. Circadian light is influenced by electric light beside the window

However, the performance of the luminaire for cases with windows G1 and G3 was sufficient to increase the percentage of timesteps above the threshold of 75%. Regarding cases with windows G7 and G9, using the luminaire increased the indoor m-EDI by 426.35% and 161.76% relative to the case lighting solo with window; the enhanced conditions were not above 75%, which shows the necessity of selecting the appropriate type of luminaire according to the window performance.

While the impact of windows on the inlet circadian light is a new topic that requires further investigation, considering these thirteen windows' performances, it is evident that available window properties do not accurately indicate circadian performance. Therefore, there should be more precise indicators for windows to evaluate the window circadian transmittance. Equation 1 is the equation proposed by (Hraska 2015) to calculate window circadian transmittance; by implementing the chosen windows' SPD in this equation, the circadian transmittance of the windows was calculated and listed in Table 1.

$$T_c = \frac{\sum_{380}^{580} D_\lambda . T(\lambda) . C(\lambda) . \Delta\lambda}{\sum_{380}^{580} D_\lambda . C(\lambda) . \Delta\lambda} \tag{1}$$

Figure 7 presents the performance of the windows for transmitting the circadian light regarding their Tvis and Tc. As previously mentioned, models with Windows G2, G4, G5, G6, G12, and G13 do meet the threshold above 75% of the year. Although a higher number of cases with different types of windows would give a solid understanding, according to the selected windows, the window with Tvis above 0.5 meets the circadian lighting requirements. For the cases with Tvis below this value, however, the higher Tc aids in meeting the requirement. As in the case with windows G1 and G2, while for a model with G1, Tvis, and Tc are 0.361 and 0.37, respectively, the model provides a healthy indoor for 22.55% of the year by increasing the Tc from 0.37 to 0.75 with same Tvis (G2), the percentage increases to 97.87%. By comparing the models as pairs with the identical Tvis, it is evident that an increment of Tc enhances the model's ability to meet the circadian light requirement. However, there is no linear correlation between the Tc and windows' circadian performance.

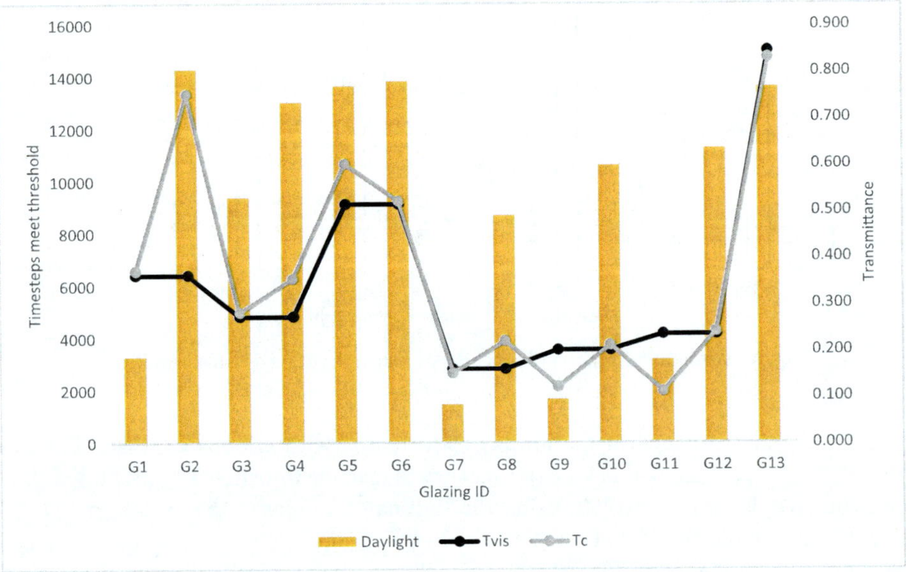

Fig. 7. Window's circadian performance regarding their Tvis and Tc

4 Conclusion

The topic of circadian light is a newly developed topic that requires further attention and exploration. So far, the impact of window visible transmittance, color, and orientation on the inlet circadian light has been discussed in different publications; however, no comprehensive study explicitly defines the impact factor of the window's properties on the inlet circadian light. This study shows the impact of the window on the inlet circadian light and how the different types of windows, even with an acceptable level of Tvis, could lead to the use of artificial light to fulfill the shortage of windows for meeting the circadian health standard. Neither Tvis nor Tc does not represent the accurate value for the circadian light transmittance and precise metrics for evaluation of the circadian light transmittance are demanded.

Acknowledgment. Research reported in this work was supported by the National Institutes of Health / National Institute of Aging #1R21AG078544–01 The Effect of Smart Ambient

Bright Light for Nursing Home Residents with Alzheimer's Disease and Related Dementias, and the National Science Foundation #2001207 CAREER: Understanding the Thermal and Optical Behaviors of the Near Infrared (NIR)-Selective Dynamic Glazing Structures.

References

Chaiyapinunt, S., Khamporn, N.: Selecting glass window with film for buildings in a hot climate. Eng. J. **13**(1), 29–42 (2009)

Circadian Lighting Design. Retrieved from WELL v2, Q4 2022 (2022). https://v2.wellcertified.com/en/wellv2/light/feature/3

Dodge, R., Daly, A., Huyton, J., & Sanders, L. (2012). The challenge of defining wellbeing. Int. J. Wellbeing **2**, 222–235 (2012)

Esaki, Y., Obayashi, K., Saeki, K., Fujita, K., Iwata, N., Kitajima, T.: Habitual light exposure and circadian activitiy rhythms in bipolar disorder: a cross-sectional analysis of the APPLE cohort. J. Affect. Disord. **323**, 762–769 (2023)

Feng, Y., Duan, Q., Wang, J., Baur, S.: Approximation of building window properties using in situ measurements. Build. Environ. **169** (2020)

Hartman, J.H.: Assessing photopic and melanopic daylighting performance of special glazing system: a full-scale experiments. Sel. Sci. Papers – J. Civil Eng. **17** (2022)

He, Q., Thomas Ng, S., Hossain, M., Skitmore, M.: Energy-efficient window retrofit for high-rise residential buildings in different climatic zones of China. Sustainability **11** (2019)

Houser, K., Esposito, T.: Human-centric lighting: Foundational considerations and a five-step design process. Front. Neurol. **12** (2021)

Hraska, J.: Chronobiological aspects of green buildings daylighting. Renew. Energy **73**, 109–114 (2015). https://doi.org/10.1016/j.renene.2014.06.008

Huizenga, C., Zhang, H., Mattelaer, P., Yu, T., Arens, E.: Window performance for human thermal comfort. Center Built Environ. (2006)

Acosta, I., Campano, M.Á., Leslie, R., Radetsky, L.: Daylighting design for healthy environments: analysis of educational spaces for optimal circadian stimulus. Sol. Energy **193**, 548–596 (2019)

Acosta, I., Leslie, R.P., Figueiro, M.G.: Analysis of circadian stimulus allowed by daylighting in hospital rooms. Light. Res. Technol. **49**, 49–61 (2017)

Jao, Y.-L., et al.: Effect of ambient bright light on behavioral and psychological symptoms in people with dementia: a systematic review. Innov. Aging **6** (2022)

Potočnik, J., Košir, M.: Influence of geometrical and optical building parameters on the circadian daylighting of an office. J. Build. Eng. **42** (2021)

Sun, Y., Wilson, R., Wu, Y.: A review of transparent insulation material (TIM) for building energy saving and daylight comfort. Appl. Energy **226**, 713–729 (2018)

Chen, X., Zhang, X.: Glazing type (colour and transmittance), daylighting, and human performances at a workspace: a full-scale experiment in Beijing. Build. Environ. **153**, 168–185 (2019)

Xiao, H., Cai , H., Li, X.: Non-visual effects of indoor light environment on humans: a review. Physiol. Behav. (2021)

Yamakawa, M., Tsujimura, S.-i., Okajima, K.: A quantitative analysis of the contribution of melanopsin to brightness perception. Sci. Rep. **9** (2019)

Yun, S.-I., Jeong, J.-W., Choi, A.: Photopic illuminance-based black-box model for regulation of human circadian rhythm via daylight control. Build. Environ. **203**, 108069 (2021)

Zhang, C., Zhu, Z., Zhao, J., Li, Y., Zhang, Z., Zheng, Y.: Ubiquitous light-emitting diodes: potential threats to retinal circadian rhythms and refractive development. Sci. Total Environ. **862**, 160809 (2023)

Smart Grids from Nano to Macro

Emerging Opportunities in Low-Frequency Variability of Renewable Resources: A 7-Year Update

Nir Y. Krakauer[1,2(✉)]

[1] Department of Civil Engineering, The City College of New York, New York, NY 10031, USA
nkrakauer@ccny.cuny.edu

[2] Earth and Environmental Sciences, City University of New York Graduate Center, New York, NY 10016, USA
https://nirkrakauer.net/

Abstract. Some recent developments in quantifying, applying, and forecasting variability in renewable resources (especially solar and wind) at timescales of months to years are reviewed. These include detailed studies of national-level resource variability, methodological work on quantifying variability and applying risk assessment to optimize and manage power grids, and work on the potential and benefits of long-range forecasting. Likely directions for future progress include data fusion, nonparametric statistics, machine learning, and joint consideration of multiple supply, demand, and storage modalities.

Keywords: resource assessment · solar energy · wind energy · value at risk · seasonal forecasting · machine learning

1 Introduction

Published in summer 2017 and based on work presented at the 2016 ASES conference, the paper "Interannual variability and seasonal predictability of wind and solar resources" [1] systematically examined interannual variability in a global reanalysis (MERRA-2) of grid-scale monthly-mean solar and wind resource (operationalized as surface incident shortwave flux and wind speed at 50-m height, respectively). Its goals were to quantify this meteorological interannual variability in renewable solar and wind resources, considering also whether anomalies in wind and solar show positive or negative correlations, and to correlate this variability with major global climate modes that might offer an avenue to forecasting it.

This research extended a previously published case study of a multi-month period of low windspeed which affected power generation over the western and southern United States in the first half of 2015, and which was associated with a Northern Pacific Mode of warming in the offshore Pacific Ocean (but not clearly

D. Renné et al. (Eds.): ASES SOLAR 2023, SPE, pp. 55–60, 2023.
https://doi.org/10.1007/978-3-031-39147-7_7

related to El Niño, which shows little correlation with windspeeds in the affected area) [2]. This event featured wind speeds regionally up to 20% below average over a multi-month period, motivating a proposed research agenda of verifying the patterns seen in this global data locally under particular site configurations and networks and developing and quantifying the value of options for acting on seasonal energy forecasts that could provide early warnings of such marked departures from the climatological renewable energy resource.

Seven years after the initial ASES presentation, the opportunity is taken here to review some of the applications and extensions of this work in the years that followed, as reflected in peer-reviewed journals, conference proceedings, and published theses. Finding recent research that has taken this direction began with the works citing [1] according to Google Scholar, augmenting these in turn with the most relevant materials cited by or citing them. While this will not cover all advances in the field, the sampling is meant to provide a sense of the main directions and extents of progress in recent years and provide the basis to identify gaps and promising directions for more detailed investigation.

2 Review of Selected Work

2.1 Quantification of Variability

One area of progress in recent years has been in detailed national-level studies of solar or wind variability. For example, [3] consider the interannual coefficient of variation of both global horizontal irradiance and direct normal irradiance across North and Central America using the National Renewable Energy Laboratory (NREL) satellite-derived National Solar Radiation Database (NSRDB). Interannual variability is relatively larger for normal as compared to global irradiance, and, similar to [1], the Midwest and Northwest regions of the United States had particularly high interannual variability in solar radiation in the winter months. [4] map the coefficient of variation of mean annual windspeed over India using the new reanalysis ERA5 [5]. [6] studies interannual variability in annual mean windspeed data collected for roughly one decade at different heights from weather stations spread across South Africa as part of the Wind Atlas for South Africa project. The author found significant increasing windspeed trends at two of the stations. [7] studies interannual variability in monthly power production of wind farms across the United States.

2.2 Methodological Advances

Methodological work on how to best quantify interannual variability remains important, and a number of noteworthy contributions can be highlighted. As part of the first author's doctoral work at the University of Colorado, [8] compare measures of interannual variability in MERRA-2 windspeed to examine which is the best predictor of variations in actual monthly power generation at wind farms across the United States, recommending the use of a robust coefficient of variation that uses the mean absolute deviation and the median instead of the standard deviation and mean. They find that this measure requires around 10 years of windspeed data to be estimated with 90% confidence to within 10%. Similarly, [9] study wind data at 2 sites in Brazil to estimate the effect of longer monitoring period in decreasing uncertainty related to power generation at a given site.

One area of methodological development has been in assessing the complementarity of multiple renewable energy resources, particularly wind and solar, from the standpoint of the coherence of their low-frequency variability. In an ISES Solar World Congress 2019 paper, [11] analyze the variance spectrum across timescales from hours to years of wind power data alongside simulated solar generation in Minnesota state, finding that the combination of wind and solar has lower variance at daily and yearly scales due to stronger wind at night and winter when solar is unavailable. A study for Texas state [12] also finds complementarity between wind and solar generation, along with favorable effects of geographic dispersion of the generating facilities. In Germany, on the other hand, complementarity between wind and solar resource, as assessed by fitting a bivariate copula, cannot overcome the large interannual variability seen for wind, and within-country geographic diversification doesn't help much [13].

2.3 Energy System Analyses

Another category of studies considers variability in solar and wind resources as part of an analysis of particular power grids, including their effect on the sizing and management of storage and fuel backups. In an IEEE Power & Energy Society conference paper, [14] present a discrete convolution approach that uses empirical linear quantile models to relate wind, sun, and energy demand in an isolated Australian minigrid to quantify the probabilities of system failure, concluding that the share of renewables could be increased while reducing dependence on fossil fuel backup for greatly reduced cost. Also as part of the same first author's Stanford doctoral dissertation, [15] consider optimal combinations of energy generation and storage for maintaining specified reliability levels in the New England power grid, again using historical data to estimate models for correlations between, for example, wind and solar generation. [16] simulate 1000 years of California electricity supply and wholesale price using stochastic weather inputs in the California and West Coast Power System (CAPOW) computer model. In their simulations, highest prices were associated with hot and dry years that have high energy demand and low hydroelectricity generation, though this could presumably be overcome by additional large solar deployments and diurnal storage that would supply ample energy under sunny summer conditions. In a master's thesis, [17] considers sub-hourly wind and solar data from generation facilities in Nova Scotia province, for which both wind and solar generation showed lower capacity factors during the winter peak demand period, finding that wind generation plus battery storage seemed more economical for reducing fossil fuel backup generation than solar generation, but highlighting the need for additional years of data to better understand meteorological variability.

Moreover, an International Conference on Machine Learning paper [18] discusses the development of a long short-term memory neural network regression model for estimating United States regional electricity demand as a function of temperature to facilitate studying meteorological factors affecting energy demand alongside supply, finding that the relationship between electricity demand and temperature is stronger in summer than in winter.

2.4 Long-Range Forecasting

Particularly intriguing is emerging work on the extent to which seasonal forecasting [19,20] could aid in managing renewable-based power systems. [21] show a geographically complex relationship between climate modes of variability, particularly the North Atlantic Oscillation (NAO), and winter sunshine over the British Isles, with positive NAO being associated with more cloud on the western side of both Britain and Ireland but less cloud in their eastern sides. [22], part of the first author's doctoral thesis, considered risk of winter wind generation shortfall during the daily evening consumption peak in France, finding that this risk is correlated with large-scale atmospheric circulation patterns and that it can be quantified with some skill at least one month ahead based on weather models. More comprehensive work on the benefits of forecasting to energy networks should include multiple renewable generation modes (wind, solar, hydroelectric, etc.) as well as demand (including its dynamic management at timescales of minutes to months) and various forms of storage.

3 Future Prospects

This unsystematic overview serves to illustrate the continuing interest in understanding low-frequency variability in solar and wind resource and its predictability. While the research summarized here has helped quantify the magnitude of interannual variablity and provided indications that it can be forecast at least a few weeks ahead, there is still a long way to go in quantifying the risks of large difference from normal in wind or solar over a given season and in providing early warning of such an event to minimize its impact on system reliability and cost. Progress in the coming years will include comparing and improving data on renewable resources (combining stations, satellites, and reanalyses); advances in statistical and machine learning methods of quantifying variability and correlation; predictability improvements from both dynamic and statistical seasonal forecast systems; and more sophisticated risk assessment amid a broader range of options and management strategies, which will also help better quantify the benefits of this research direction to different communities.

References

1. Krakauer, N.Y., Cohan, D.S.: Interannual variability and seasonal predictability of wind and solar resources. Resources **6**, 29 (2017). https://doi.org/10.3390/resources6030029

2. Rife D.; Krakauer N.Y.; Cohan D.S. , Collier, J.C.: A new kind of drought: U.S. record low windiness in 2015. Earthzine, IEEE (2016). https://earthzine.org/a-new-kind-of-drought-u-s-record-low-windiness-in-2015/
3. Habte, A., Sengupta, M., Gueymard, C., Golnas, A., Xie, Y.: Long-term spatial and temporal solar resource variability over America using the NSRDB version 3 (1998–2017). Renew. Sustain. Energy Rev. **134**, 110285 (2020). https://doi.org/10.1016/j.rser.2020.110285
4. Bastin, J., Katyal, R., Kumar, R.V., Lakshmi, P.Y.: Inter Annual Variability of wind speed in India. Int. J. Ambient Energy Informa **43**, 5232–5246 (2021). https://doi.org/10.1080/01430750.2021.1945492
5. Hersbach, H., et al.: Operational global reanalysis: progress, future directions and synergies with NWP. ECMWF technical report (2018). https://doi.org/10.21957/tkic6g3wm
6. Chauke, M.: Trend analysis and inter-annual variability in wind speed in South Africa. J. Energy in South Afr Acad. Sci. South Afr. **33**, 13–21 (2023). https://doi.org/10.17159/2413-3051/2022/v33i4a13162
7. Boretti, A., Castelletto, S.: Low-frequency wind energy variability in the continental contiguous United States. Energies **13**, 144 (2019). https://doi.org/10.3390/en13010144
8. Lee, J.C.Y., Fields, M.J., Lundquist, J.K.: Assessing variability of wind speed: comparison and validation of 27 methodologies. Wind Energy Sci. Discuss. **3**, 845–868 (2018). https://doi.org/10.5194/wes-3-845-2018
9. Miguel, J.V.P., Fadigas, E.A., Sauer, I.L.: The influence of the wind measurement campaign duration on a measure-correlate-predict (MCP)-based wind resource assessment. Energies **12**, 3606 (2019). https://doi.org/10.3390/en12193606
10. Ghosh, R.; Craig, M.; Matthews, H.S.; Samaras, C., Berti-Equille, L.: Reconstruction of long-term historical demand data. arXiv (2022). https://doi.org/10.48550/ARXIV.2209.04693
11. Perez, M., Perez, R.: On the complementary variability of wind and solar power. In: Proceedings of the ISES Solar World Congress 2019, International Solar Energy Society (2019). https://doi.org/10.18086/swc.2019.42.06
12. Slusarewicz, J.H., Cohan, D.S.: Assessing solar and wind complementarity in Texas. Renew. Wind Water Solar **5**(1), 1–13 (2018). https://doi.org/10.1186/s40807-018-0054-3
13. Schindler, D., Behr, H.D., Jung, C.: On the spatiotemporal variability and potential of complementarity of wind and solar resources. Energy Conver. Manag. **218**, 113016 (2020). https://doi.org/10.1016/j.enconman.2020.113016
14. Gao, W.; Tayal, D., Gorinevsky, D.: Probabilistic planning of minigrid with renewables and storage in Western Australia. In: 2019 IEEE Power & Energy Society General Meeting (PESGM) (2019). https://doi.org/10.1109/pesgm40551.2019.8973483
15. Gao, W., Gorinevsky, D.: Probabilistic modeling for optimization of resource mix with variable generation and storage. In: IEEE Transactions on Power Systems, Institute of Electrical and Electronics Engineers (IEEE), 35: 4036–4045 (2020). https://doi.org/10.1109/tpwrs.2020.2984492
16. Su, Y., Kern, J.D., Reed, P.M., Characklis, G.W.: Compound hydrometeorological extremes across multiple timescales drive volatility in California electricity market prices and emissions. Appl. Energy **276**, 115541 (2020). https://doi.org/10.1016/j.apenergy.2020.115541

17. Kiefte, D.: Using Energy Storage to Maximize Wind and Solar Electricity Generators Effectiveness in Dispatchable Generation Replacement. Dalhousie University (2021). http://hdl.handle.net/10222/80788
18. Ghosh, R., Craig, M., Matthews, H.S., Samaras, C., Berti-Equille, L.: Reconstruction of long-term historical demand data. arXiv (2022). https://doi.org/10.48550/ARXIV.2209.04693
19. Krakauer, N.Y., Grossberg, M.D., Gladkova, I., Aizenman, H.: Information content of seasonal forecasts in a changing climate. Adv. Meteorol. Hindawi **2013**, 480210 (2013). https://doi.org/10.1155/2013/480210
20. Krakauer, N.Y.: Temperature trends and prediction skill in NMME seasonal forecasts. Clim. Dyn. **53**(12), 7201–7213 (2017). https://doi.org/10.1007/s00382-017-3657-2
21. Correia, J. M., McDermott, F., Sweeney, C., Doddy, E., Griffin, S.: An investigation of the regional correlation gradients between Euro-Atlantic atmospheric teleconnections and winter solar short wave radiation in northwest Europe. Meteorol. Appl. **27** (2020). https://doi.org/10.1002/met.1892
22. Alonzo, B., Drobinski, P., Plougonven, R., Tankov, P.: Measuring the risk of supply and demand imbalance at the monthly to seasonal scale in France. Energies **13**, 4888 (2020). https://doi.org/10.3390/en13184888

Applications of PV and Solar Thermal

Experimental Study of a Small Beam-Down and Direct-Down Solar System for Water Preheating/Desalination Applications

Erick Moreno Resendiz(✉), Javed Akhtar, Krishna Kota, and Sarada Kuravi

Department of Mechanical and Aerospace Engineering, New Mexico State University, 1040 S. Horseshoe Street, Las Cruces 88003, USA
emorenor@nmsu.edu

Abstract. The feasibility of multiple scalable direct solar configurations for small-scale water preheating/desalination was assessed experimentally and complemented using a theoretical model. The physics of the beam-down optics, direct-down optics, and a combination of the two showed to satisfactorily enhance the water yield by 11, 28, and 25% respectively due to augmented heat radiation into the basin. In addition, these configurations benefit from the increased heat transfer coefficient by allowing a small water depth over a relatively large area.

Keywords: Solar Thermal Energy · Desalination · Beam-Down Solar Concentrator · Solar Heating · Low-Temperature Heating

1 Introduction

Water scarcity is a significant global issue, water consumption has increased by 600% over the past century, and a 55% increase in demand is projected by 2050 [1, 2]. In 2015, around 663 million people lacked access to clean drinking water [3].

Desalination is a potential solution, but its high energy needs are challenging. Currently, 99% of desalination processes rely on fossil fuels [4]. Researchers are exploring renewable energy options, with solar energy being a promising choice in arid regions [5].

Solar desalination can use solar thermal energy (STE) indirectly or directly. Indirect systems use solar energy to generate electricity for the desalination processes [6, 7], but they are energy-intensive and require large land areas for photovoltaic panels. Direct systems, like solar stills (SS), simplify the process by using the STE directly for distillation [8, 9].

Researchers have made efforts to improve direct solar desalination systems by modifying SS structures, incorporating thermal energy storage materials, altering the interfacial evaporation [10–12]. Notably, the utilization of concentrated solar power (CSP) has the potential to unlock the full potential of direct solar desalination, making it a more viable and scalable solution [13–16].

The objective of this study is to assess the viability of two highly scalable concepts, namely Beam-Down (BDC) and Direct-Down (DDC) Solar Concentration, for the purpose of water preheating and desalination applications.

D. Renné et al. (Eds.): ASES SOLAR 2023, SPE, pp. 63–71, 2023.
https://doi.org/10.1007/978-3-031-39147-7_8

2 Materials

Figure 1 depicts the experimental setup used in the study, consisting of 11 heliostats arranged around a testing basin, a water level controller, and a central mirror for Beam-Down Concentration (BDC) testing.

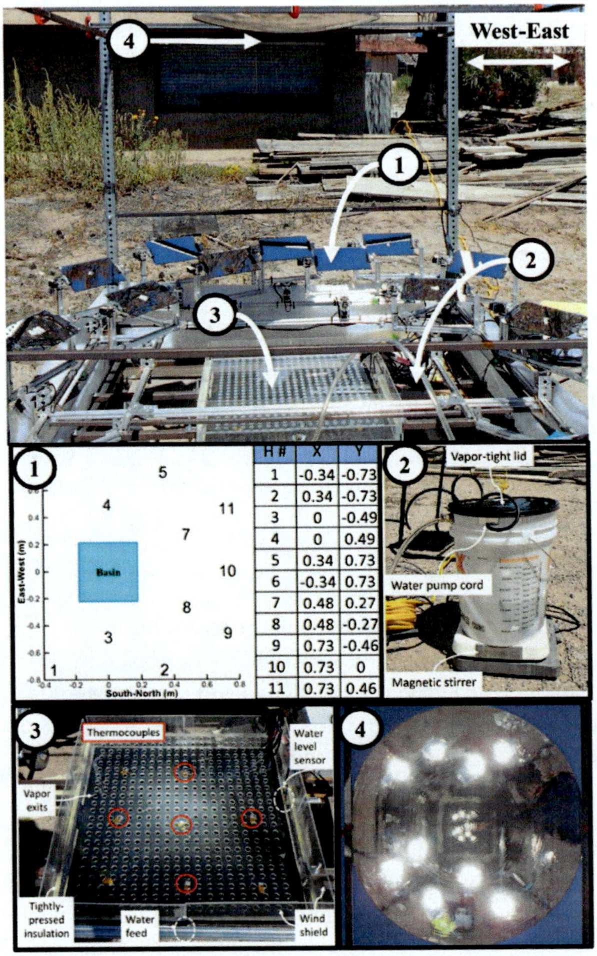

Fig. 1. Main components of the experimental setup. 1) Heliostat field distribution, 2) Water level controller, 3) Polycarbonate basin and 4) circular hyperbolic central mirror (for the BDC configuration).

3 Methods

Experiments were conducted from 9 AM to 6 PM at the New Mexico State University (NMSU) Southwest Technology Development Institute (SWTDI) in Las Cruces, New Mexico (Latitude: 32°16′14.9″N, Longitude: 106°44′44.6″W). The measurements included 5 surface temperatures (T_s), solar irradiation (G_o), ambient temperature (T_∞), and windspeed (V_∞). K-type thermocouples and a Ni cDAQ-9174 data acquisition system controlled via LabVIEW software were used for temperature measurements, while a SPM-1116SD solar meter and anemometer were employed for solar irradiation and windspeed readings, respectively. The net water yield was determined by measuring the initial and final evaporated volume of saline water (35 gr/kg). The experiments were carried out in an insulated polycarbonate basin with controlled water depth (~1.75 cm) on black-painted 6061 aluminum plates. Three configurations were tested: baseline (BL), Direct-Down concentration (DDC), and Beam-Down solar concentrator (BDC), with the hybrid DDC + BDC configuration also explored (see Fig. 2). Additionally, an in-house model described in [17] was used to analyze further the DDC and BDC configurations based on cosine efficiency and solar irradiation data. The experimental setup and the in-house model illustration under the DDC configuration at the same timeframe is shown in Fig. 3.

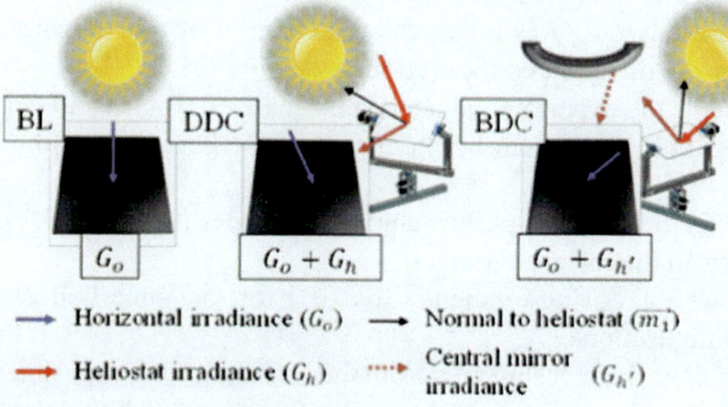

Fig. 2. Different configurations tested for water desalination.

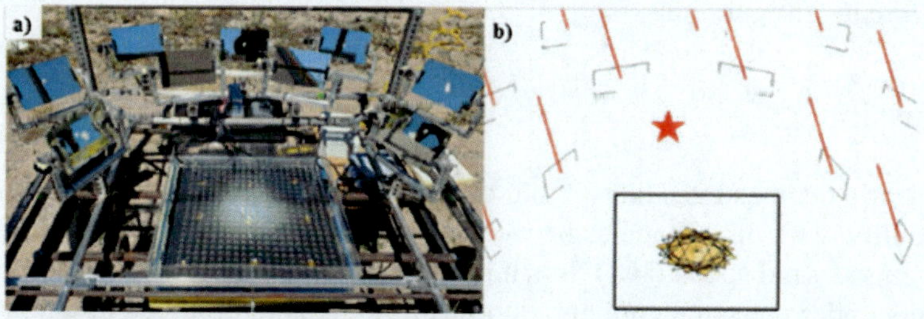

Fig. 3. DDC configuration experimental and model at the same timeframe, corresponding to September 5th, 2022, at 13:35 h.

4 Results

The characterization of the 8 days in which experiments were conducted is presented in Fig. 4.

Figure 4a-b illustrate the variation of available irradiance (G_o) and the variation of ambient temperature (T_∞) throughout the experiment duration. As anticipated, the irradiance and temperature increase as the dates approach the summer solstice, resulting in higher values for experiments conducted closer to the summer solstice.

Figure 4.c demonstrates the local instantaneous wind speed (V_∞) at the experimental site. Unlike G_o and T_∞, V_∞ is not directly linked to the date and is more challenging to homogenize. Accurate characterization of V_∞ helps explain sudden changes in temperature due to variations in convective heat loss.

The results of the 8 experiments (4 configurations repeated twice) are shown in Fig. 5.

Figure 5.a shows the average temperature ($T_{s,avg}$) calculated from the five measured surface temperatures. Figure 5.b displays the maximum temperature ($T_{s,max}$) among the five surface temperatures, typically corresponding to the central temperature. The contrast between these two figures provides an indication of the temperature variation within the basin for the different configurations.

Figure 5.c presents the experimental net and normalized water yield per unit area for the different configurations, comparing them against the best-performing baseline run (BL#1).

To further explain the obtained performance of the experimental data, Fig. 6 provides additional results from the in-house model.

Figure 6.a shows the cosine efficiency ($cos(\theta)$) for the same heliostats under the DDC and BDC configurations.

Figure 6.b combines the measured G_o from the corresponding experimental day with the theoretical cosine efficiency and zenith angle (θ_s) to establish the global horizontal irradiance (G_h). For the BDC configuration, G_h was multiplied by the reflectivity of the central mirror (ρ_2). The expected irradiances after the heliostat reflection were compared for the different configurations.

5 Discussion

Figure 4 clearly demonstrates that all four different configurations exhibit a certain level of repeatability, with discrepancies between repeated experiments mainly attributed to variations in the wind speed (V_∞). In simpler terms, larger differences in V_∞ between experiments under the same configuration result in greater variations in temperatures, including the average surface temperature ($T_{s,avg}$). Nonetheless, the overall trends remain consistent, indicating that the hybrid configuration (DDC + BDC) outperforms the others, even when solar irradiance is lower (as depicted in Fig. 4.a).

Comparing the DDC and BDC configurations individually, both have their drawbacks, which are mitigated when combined. The DDC configuration is not significantly affected by the inclusion of the second reflection (lower optical efficiency due to imperfect reflectivity). However, the required positioning leads to more abrupt changes in cosine efficiency, shortening the system's full-operating window. Specifically, when the

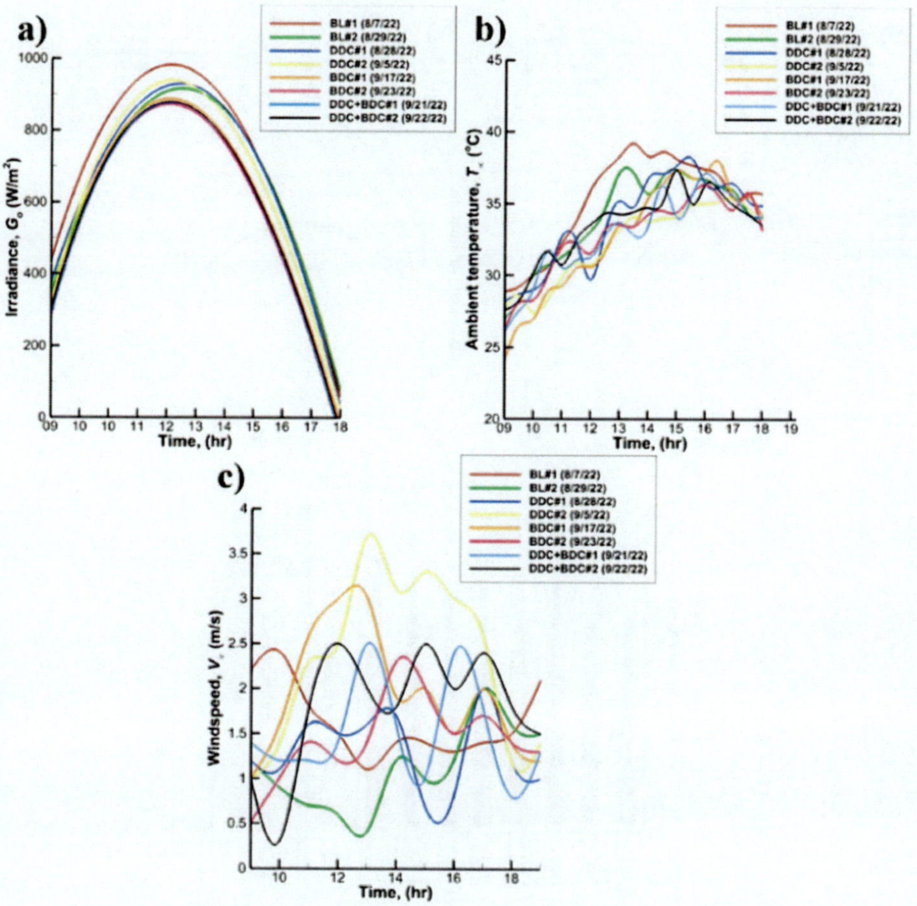

Fig. 4. Characterization of days in which experiments were conducted.

sun is near horizon (early morning or late evening), heliostats in the same semicircle as the sun (East or West semicircles) become less effective, while those in the opposite semicircle perform better (see Fig. 1).

In contrast, the BDC configuration provides greater stability but is susceptible to the impact of imperfect reflectivity in the central mirror. However, by incorporating less efficient heliostats from the DDC configuration into the BDC setup, an enhanced performance is achieved, as evidenced in the hybrid: DDC + BDC configurations.

The maximum temperature in the basin, also shows the superiority of the hybrid configuration (see Fig. 5.b). Regardless of the lower -and lesser availability of- G_o, the hybrid configuration still showed a higher $T_{s,max}$ compared to its separate constituents (DDC and BDC). The same explication provided for Fig. 5.a is presumed. Because the DDC showed more satisfactory results (in general), the hybrid configuration was tended more toward a DDC with just slight implementations of the BDC. The change of the heliostats from DDC to BDC (or vice versa) was done empirically; this in combination

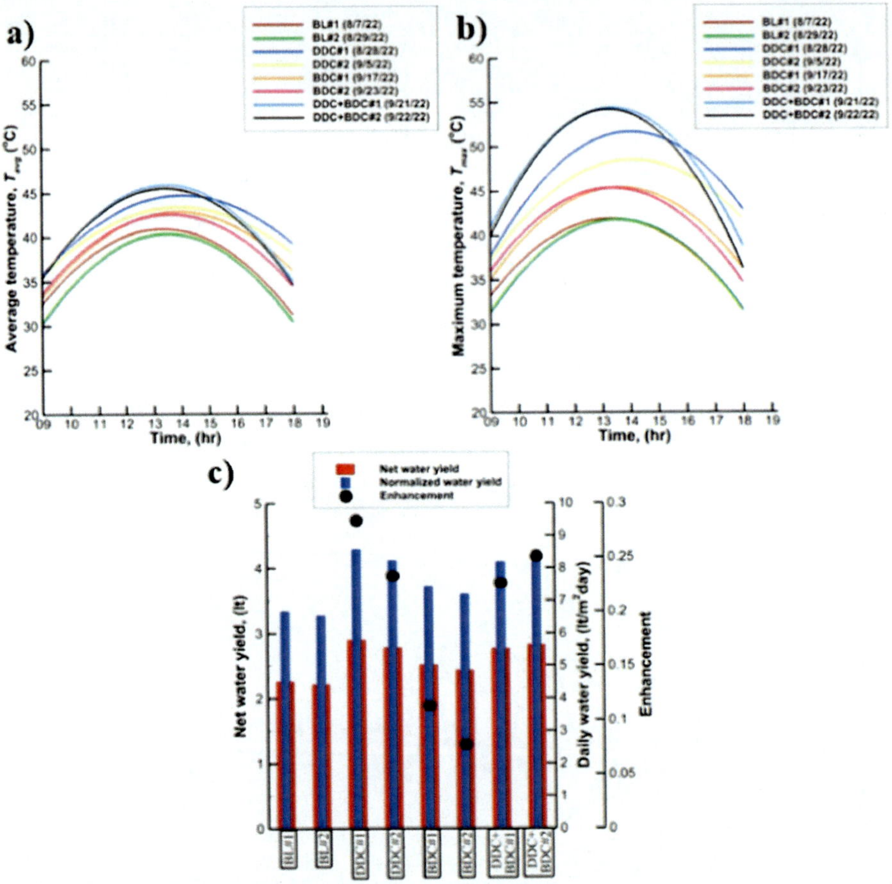

Fig. 5. Experimental results for the tested configurations.

with the shorter daytime of the experiment day account for the sudden decrease later in the day. Having this change automated, depending on a threshold would have granted (at least) a comparable temperature to the DDC.

The water yield is in accordance with both the $T_{s,avg}$ and $T_{s,max}$. The higher and prolongated G_o resulted in a higher water yield for both the DDC and the hybrid configurations. It is presumed that having automated the change in the hybrid configuration would have resulted in a higher water yield compared to the DDC alone.

The hybrid, the DDC and the BDC configurations showed an enhancement of 25, 28 and 11%, respectively when compared against the best BL result (BL#1). However, it must be noted that the low performance of the BDC is attributed to the low reflectivity of the central mirror (ρ_2), as the average cosine efficiency for this configuration is higher; and to the lower collection efficiency (ε_c), caused by the poor cohesion of the central mirror-basin-heliostat. In other words, due to the lower ε_c (high magnification causes the image to be outside the basin) and the inclusion of the ρ_2 the available energy from the BDC configuration was lower than that of the DDC regardless of the higher

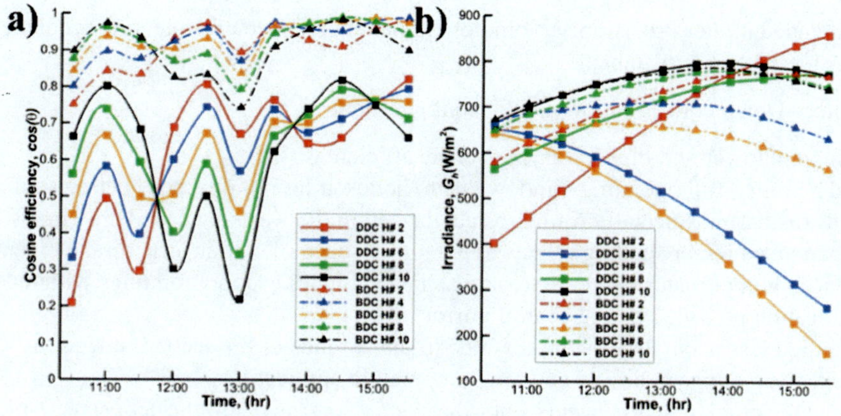

Fig. 6. Theoretical model cosine efficiency and heliostat irradiance.

and stabler cosine efficiency (assuming perfect heliostat's reflectivity ($\rho_1 = 1$) for both configurations) (see Fig. 6). This resulted in a lower rate of heat (q), and subsequently, a lower available heat (Q).

In addition, the tendency of the DDC to concentrate, (or have very low magnifications) provided a higher irradiance at the basin explaining the higher values for $T_{s,max}$. The evaporation rate, being a function of the temperature, increased with the higher $T_{s,max}$.

However, due to the wrong cohesion of the central mirror-basin-heliostat, the lower collection efficiency (ε_c) of the BDC configuration (would yield to a lower available rate of heat (q) whereas the DDC configuration has a perfect collection at the analyzed high-intensity window; thus, justifying the better performance of the DDC configuration.

6 Conclusions

The present study successfully tested and compared three different configurations utilizing direct solar thermal energy for water desalination. The results demonstrated significant improvements in water yield and surface temperature. However, certain limitations were observed for each configuration.

Beam-Down solar concentrator (BDC) configuration:

1. Showed higher and less varying cosine efficiency.
2. Operated throughout the day with the exception of shading constraints.
3. Experienced additional shading effects that hindered performance.
4. Required a minimum central mirror reflectivity for inclusion.
5. Magnified irradiance did not significantly increase water temperature or achieve high evaporation rates compared to Direct-Down concentration (DDC).
6. Emphasized the importance of cohesion between the central mirror, basin, and heliostat to maximize useful heat and collection efficiency.
7. Enabled more design options and applications through the inclusion of a second reflection.

8. Demanded higher precision servomotors to maintain expected collection efficiency.
9. Considered easily scalable.

Direct-Down concentration (DDC) configuration:

1. Exhibited lower and highly varying cosine efficiency.
2. Had a limited full operating window due to heliostat limitations in reflecting sunlight towards the basin, especially for heliostats located in the west and east demi circles.
3. Experienced lesser shading effects due to the absence of certain structural elements.
4. Induced higher water temperature and achieved higher evaporation rates without the magnification provided by the central mirror.
5. Required less accurate servomotors due to the absence of the second reflection.
6. Lacked versatility in design and low-temperature applications.
7. Suited for smaller-scale systems when the basin is located below the heliostats' z-plane and all heliostats are in the same z-plane.

Hybrid DDC + BDC configuration:

1. Proved advantageous when central mirror reflection was not very high.
2. Offered potential to mitigate the limitations of individual configurations, such as the limited full operating window of DDC and the inclusion of the imperfect second reflectivity in BDC.
3. Required careful selection of heliostats based on cosine efficiencies and central mirror reflectivity.
4. Enabled the use of larger basins with regions of higher evaporation rates by combining the "high" irradiance over a smaller area (DDC configuration) with the magnified irradiance over a larger area (BDC).

In conclusion, this study provides valuable insights into the performance and characteristics of the BDC, DDC, and hybrid DDC + BDC configurations for direct solar thermal water desalination applications. These findings can inform the development of optimized and efficient systems for various scale applications in the future.

Acknowledgments. This project is supported by the US Bureau of Reclamation, Award Number R19AC00110. The authors wish to extend their gratitude to M.S. Umar Masood for contributing to the development and assembly of the control system and to Dr. Olga Lavrova for kindly providing us with a facility to conduct the experiments.

References

1. T. United, N. World, and W. Development, WATER AND
2. Schmidt, J.J.: Valuing water (2019). https://doi.org/10.4324/9780429453571-2
3. 2015 Update and MDG Assessment (2015)
4. Boretti, A., Rosa, L.: Reassessing the projections of the World Water Development Report. NPJ Clean Water **2**(1), 15 (2019). https://doi.org/10.1038/s41545-019-0039-9
5. Chauhan, V.K., Shukla, S.K., Tirkey, J.V., Singh Rathore, P.K.: A comprehensive review of direct solar desalination techniques and its advancements, vol. 284. Elsevier Ltd., Amsterdam (2021). https://doi.org/10.1016/j.jclepro.2020.124719

6. Gross, A., Morvezen, A., Gomez, P.C., Xu, X., Xu, P.: Numerical investigation of the effect of two-dimensional surface waviness on the current density of ion-selective membranes for electrodialysis. Water (Switzerland) **11**(7), 1397 (2019). https://doi.org/10.3390/w11071397
7. Chen, L., Xu, P., Wang, H.: Interplay of the factors affecting water flux and salt rejection in membrane distillation: a state-of-the-art critical review. Water (Switzerland) **12**(10), 2841 (2020). https://doi.org/10.3390/w12102841
8. Yong Ho, Z., Bahar, R.: Performance analysis of a solar desalination system with concentrated solar power (CSP). In: IOP Conference Series: Earth and Environmental Science, vol. 268, no. 1 (2019). https://doi.org/10.1088/1755-1315/268/1/012023
9. Manchanda, H., Kumar, M.: A comprehensive decade review and analysis on designs and performance parameters of passive solar still. Renew.: Wind Water Solar **2**(1), 1–24 (2015). https://doi.org/10.1186/s40807-015-0019-8
10. Sathyamurthy, R., Harris Samuel, D.G., Nagarajan, P.K., Arunkumar, T.: Geometrical variations in solar stills for improving the fresh water yield—a review. Desalination Water Treat **57**(45), 21145–21159 (2016). https://doi.org/10.1080/19443994.2015.1136241
11. Arunkumar, T., Vinothkumar, K., Ahsan, A., Jayaprakash, R., Kumar, S.: Experimental study on various solar still designs. ISRN Renew. Energy **2012**, 1–10 (2012). https://doi.org/10.5402/2012/569381
12. Chen, L., Xu, P., Kota, K., Kuravi, S., Wang, H.: Solar distillation of highly saline produced water using low-cost and high-performance carbon black and airlaid paper-based evaporator (CAPER). Chemosphere **269**, 129372 (2021). https://doi.org/10.1016/j.chemosphere.2020.129372
13. Voropoulos, K., Mathioulakis, E., Belessiotis, V.: Experimental investigation of a solar still coupled with solar collectors. Desalination **138**(1–3), 103–110 (2001). https://doi.org/10.1016/S0011-9164(01)00251-X
14. Gorjian, S., Ghobadian, B., Tavakkoli Hashjin, T., Banakar, A.: Experimental performance evaluation of a stand-alone point-focus parabolic solar still. Desalination **352**, 1–17 (2014). https://doi.org/10.1016/j.desal.2014.08.005
15. Fathy, M., Hassan, H., Salem Ahmed, M.: Experimental study on the effect of coupling parabolic trough collector with double slope solar still on its performance. Solar Energy **163**, 54–61 (2018). https://doi.org/10.1016/j.solener.2018.01.043
16. Mu, L., et al.: Enhancing the performance of a single-basin single-slope solar still by using Fresnel lens: experimental study. J. Clean Prod. **239**, 118094 (2019). https://doi.org/10.1016/j.jclepro.2019.118094
17. Resendiz, E.M., et al.: Analysis of a small-scale modified beam-down solar concentrator system for low-temperature applications. Renew. Energy 119007 (2023). ISSN: 0960-1481. https://doi.org/10.1016/j.renene.2023.119007, https://www.sciencedirect.com/science/article/pii/S0960148123009199

Performance Evaluation of a Floating Photovoltaic System in Saudi Arabia

Shafiqur Rehman[1(✉)], Ahmad Al Muhanna[1], Luai M. Alhems[2], Mohammed A. Mohandes[3], and Ali A. Alshaikhi[3]

[1] Interdisciplinary Research Center for Renewable Energy and Power Systems (IRC-REPS), King Fahd University of Petroleum and Minerals (KFUPM), Box 767, Dhahran, Saudi Arabia
srehman@kfupm.edu.sa

[2] Applied Research Center for Metrology, Standards, and Testing (ARC-MST), King Fahd University of Petroleum and Minerals (KFUPM), Box 767, Dhahran, Saudi Arabia

[3] Electrical Engineering Department, King Fahd University of Petroleum and Minerals (KFUPM), Box 767, Dhahran, Saudi Arabia

Abstract. This study reports the potential of Floating Photovoltaic (FPV) systems in the Saudi Arabian environment. As the temperature of PV module increases, its efficiency drops depending on its rated temperature coefficient. The solar panel back surface is mostly affected by the temperature due to heat encapsulation. The temperature difference between the ground mounted and the FPV systems reached a maximum of 16.15 °C during May. Even a moderate reduction in PV module temperature can play an important role in large scale solar power plants over a long period of time. This demonstration set up includes the design of an FPV system of 171.9 kWp capacity installed on the water surface a lake around the centre of KFUPM campus.

Keywords: Floating Photovoltaic (FPV) Plant · PV efficiency · Back Surface Temperature · Energy Output

1 Introduction

To reach the CO2 emission target set by the VISION 2030 of the Kingdom of Saudi Arabia, the share of renewables in its energy mix has to be increased significantly. Towards the set goal, the proposed system will be an excellent option for power generation at places having bounded water bodies. In the proposed modern "LINE" cities in the RED SEA areas; such systems will be extremely useful to generate power efficiently and at the same time conserve water by minimizing the surface water evaporation losses. As of 2014, there were 482 dams in the country with total capacities of 2.08 Billion cubic meters of water. Some of these dams are located in Al-Baha (22), Asir (16), Jazan (2), Madina (5), Mecca (11), Najran (1), Riyadh (6) and many others. Such dams and water reservoirs can be utilized to deploy the floating PV systems and can generate power for the local communities. As an example, Baysh dam is a gravity dam on Wadi Baysh about 35 km northeast of Baysh in Gizan, as shown in Fig. 1. Its catchment area is 4,843 km^2 and volume capacity is 675,000 m^3.

D. Renné et al. (Eds.): ASES SOLAR 2023, SPE, pp. 72–79, 2023.
https://doi.org/10.1007/978-3-031-39147-7_9

Fig. 1. The Baysh dam near Gizan

FPV is relatively a new concept, with few commercial deployments. However, proof of concept type of implementations is available around the world. To conserve the valuable land & water, installing Solar PV system on water bodies like oceans, lakes, lagoons, reservoirs, irrigation ponds, waste water treatment plants, wineries, fish farms, dams, and canals can be an attractive option [1]. Floating type solar photovoltaic panels have numerous advantages compared to ground based solar PV plants, including fewer obstacles to block sunlight, convenient, energy efficiency, and higher power generation efficiency owing to its lower temperature underneath the panels. Photovoltaic technology is the most common, effective, sustainable, and eco-friendly for direct power generation in the field of renewable energy and does not involve any moving part [2–6]. Some of the existing proof of concept implementations of FPV plants in different regions of the globe are listed in Table 1. Two of such proof of concept FPV plants are shown in Figs. 2 and 3, [7, 8]. Generally, FPV systems are similar to the regular ground or roof mounted PV ones, with the exception that the panels and the inverters are mounted on a floating structure, called floats. The DC current generated by PV modules is directed towards combiner boxes and converted to AC current by inverters. Generally, in small plants which are very close to shore, inverters can be placed on land. Some extra arrangements are necessary for floating PV compared to regular PV systems.

The primary components of a floating PV system are the following, (Fig. 4):

- **Pontoon/floats:** For installation of floating PV plant, firstly, pontoon (floating structure) is constructed which holds the solar panels and keep floating. It is a sturdy structure, which is placed on water surface carrying panels and other electrical equipment. It also provides protection from high wind velocity to a certain extent. Numerous hollow lightweight plastic floats made of fibre-reinforced plastic (FRP), high-density plastic (HDPE) or medium density plastic (MDPE) are usually used forms of a pontoon. PV panels are fitted on these pontoons maintaining suitable inclination to receive highest possible solar radiation.
- **Mooring and anchoring system:** The pontoon or floating structures are held securely by mooring and anchoring the entire system. These floating structures can freely move or fixed in a suitable position.
- **Waterproof cables and connectors:** All cables remain in and around the water and should be able to absorb the shocks and be waterproof.

Table 1. List of some of the existing FPV plants around the globe

S. No.	Installed Capacity (kW)	Location
1	175	Napa valley's Far Niente Wineries, California, USA
2	2,300	Hyogo prefecture, western Japan
3	4,000	Jamestown, South Australia
4	30	Gundlach bunshu wineries, California, USA
5	500	Bubano, Italy
6	300	Agost, Spain
7	14	Piolenc, France
8	20	Cheongju, South Korea
9	5	Bishan Park, Singapore
10	0.5	Sadbury, Canada
11	200	Sheeplands farm, Barkshire, UK
12	50,000	West Kallada, Kerala India

Fig. 2. A 10 KW FPV plant at Rajarhat in Kolkata [7]

Fig. 3. A proof of concept implementation of a 500 KW FPV plant at Wayanad in Kerala [8]

- **Lightning Protection System:** It is connected to the metal frame supporting the modules and is grounded.

- **PV Panels:** It includes PV panels, string inverter, and other electrical equipment which are essential for transmitting electricity.
- **Combiner box:** It collects the current from all the PV arrays and is connected to the central inverter.
- **Central inverter and Transformer:** The power received at central inverter is transmitted to the transmission line or the load by raising the voltage compatible with the transmission line by the transformer.

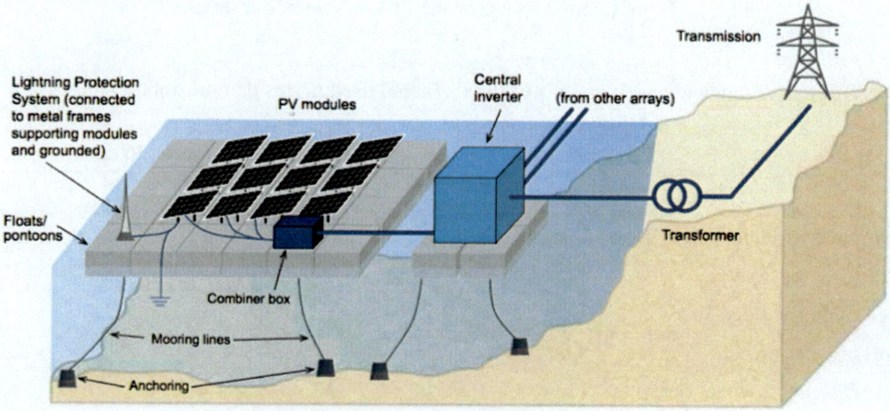

Fig. 4. Schematics of an FPV power plant [8].

This study attempts to show experimentally the potential and improved performance of floating photovoltaic (FPV) system in Saudi Arabian environment.

2 Experimental Setup Design and Fabrication

The proposed system consisted of the following equipment and sensors and the assembled system is shown in Fig. 5.

- Two monocrystalline solar panels each of 150 W.
- EPEVER charge controller.
- 100 Ah lead acid battery.
- EPEVER eLOG data logger for current and voltage logging.
- Campbell Scientific CR10X for temperature logging.
- 4 surface temperature sensors and one water temperature sensor all of K-type sensors compatible with the CR10X data logger.

The data logger and a dead load are mounted on the ground, as shown through a wooden box. The technical specifications of the PV panels are given in Table 2.

Fig. 5. Experimental set up of the FPV assembled in-house

Table 2. Technical specifications of the PV panel used in this demonstration study

Cell type	Mono crystalline silicon 156 mm × 156 mm (36 cells)
Max power	150 W
Max power voltage V_{mp}	18.62 V
Max power current I_{mp}	8.06 Amp
Open circuit voltage V_{OC}	22.33 V
Short circuit current I_{sc}	8.53 Amp
Cell efficiency	17.96
Temperature coefficient	−0.30%/C
Dimensions	(1480 × 670 × 40) mm
Weight	13 kg

The float is made up of PVC pipes using 6-inch diameter pipes (SCH 40) with 5 mm wall thickness. The float, shown in Fig. 6, is made of (2 m long and 1 m wide) pipes fitted at 90° elbows to make it waterproof. A plywood was put on the float to hold the payload consisting of scrap pieces and floor tiles to test the floatability of the load in a man-made pond located near the centre of KFUPM Campus. The proposed float was able to hold 70 kg while being 4 cm above water level. The floating test exhibited the float's effectiveness to safely hold the experimental requirement while stay floating.

Fig. 6. Fabrication and testing of proposed float designed and developed in-house

3 Experimental Results and Discussion

As mentioned earlier, the solar panel is rated with a temperature coefficient (−0.3%/°C) and in the case the highest temperature is recorded on the back side of the panel, as shown in the graph below (Fig. 7). The water temperature remains around 30 °C while the diurnal changes are recorded for both back and front surface temperature of the PV panels during 9 days measurements.

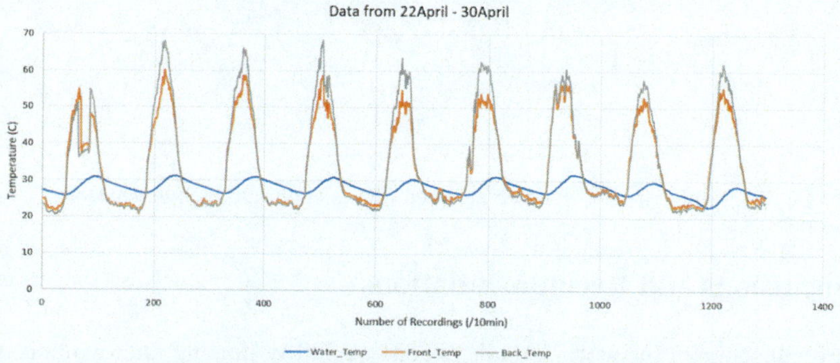

Fig. 7. 10-min averaged water, front surface, and back surface temperature during 22 to 30 April 2021.

After locating the place of the highest temperature, another panel was installed on the nearby ground to compare its temperature with the floating PV modules. The sensor was installed on the back surface of the ground mounted panel. Figure 8 compares the temperatures measured on the back surfaces of the FPV and ground mounted systems. During the entire data collection period, it is observed that the temperature measured on the back surface of the FPV system remained always less than that in case of ground mounted PV system. The maximum difference is observed on the 30th of May where the temperature of the ground mounted PV was (64.67 °C) and the FPV back surface temperature was (48.52 °C). However, on windy days the FPV system loses its advantage as both systems have almost the same temperatures.

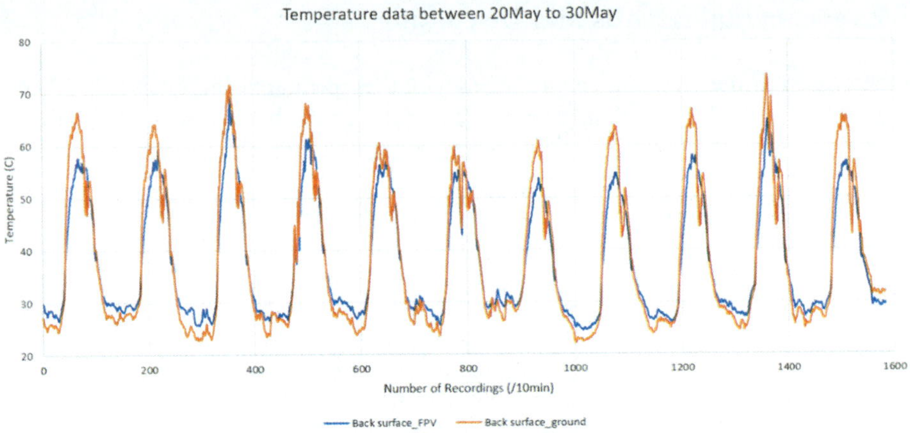

Fig. 8. Temperature difference between FPV and ground mounted system.

4 Conclusions and Recommendations

In the present scope of the work, a total capacity of 300 W floating photovoltaic system is designed, sized, assembled, and monitored in the months of July and August. The performance of both the systems FPV and ground bases is monitored by sensors installed on the back and front surfaces of the PV panels. Relatively lower temperatures are observed on the back surface of the panels of the FPV system compared to that installed on the ground surface. The temperature difference is found to be small between the two systems which could be accounted for a shallow depth of the water (only 70 cm). Such systems are expected to be more effective where the water depths are more than 2 m. The present system, though exhibited small temperature difference, might increase the power output from the FPV system. It is recommended to test such a system in water bodies with deeper water.

Acknowledgement. Authors acknowledge the support of the Renewable Energy Technical Incubator (RETI) program through project grant number CREP2522, King Fahd University of Petroleum and Minerals (KFUPM), Dhahran, Saudi Arabia.

References

1. Sahu, A., Yadav, N., Sudhakar, K.: Floating photovoltaic power plant: a review. Renew. Sustain. Energy Rev. **66**(2016), 815–824 (2016)
2. Rahman, M.M., Hasanuzzaman, M., Rahim, N.A.: Effects of various parameters on PV-module power and efficiency. Energy Convers. Manag. **103**, 348–358 (2015)
3. Pankaj, K., Sam, J.S., Shukla, A.K., Sudhakar, K., Arbind, K.: Performance analysis of 68W flexible solar PV. J. Energy Res. Environ. Technol. **2**(3), 227–231 (2015)
4. Shukla, K.N., Rangnekar, S., Sudhakar, K.: A comparative study of exergetic performance of amorphous and polycrystalline solar PV modules. Int. J. Exergy **17**(4), 433–455 (2015). https://doi.org/10.1504/IJEX.2015.071559

5. Sudhakar, K., Srivastava, T.: Energy and energy analysis of 36W solar photovoltaic module. Int. J. Ambient Energy **2**(1), 31–34 (2013)
6. https://www.energy.gov/eere/solar/solar-photovoltaic-cell-basics
7. Floating solar power plant in West Bengal, India. https://www.vikramsolar.com/case-studies/floating-solar-power-plant-in-west-bengal-india/
8. Solarplaza: Top 100 floating solar projects (2019). https://www.solarplaza.com/channels/markets/11968/top-100-floating-solar-projects/#accesstop100

Potential Electricity Production of Vertical Solar Photovoltaic Arrays

Xiaodi Hou(✉) and Juan-Carlos Baltazar

Department of Architecture, Texas A&M University, College Station, TX 77843, USA
{xdhou,juan-carlosbaltazar}@tamu.edu

Abstract. It is not hard to imagine that, in general, for electricity generation the best orientation for a tilted solar photovoltaics (PV) is facing to the true south in the north hemisphere. This study focuses on and provides the potential electricity generation for solar walls –solar PV tilted 90-degrees. The analysis is based on simulations for five cities that cover some representative climate zones of USA: Boston, College Station, Miami, Seattle and San Diego. The results showed that for vertical tilted solar walls, the best orientation of electricity production is not always facing to the south. In Miami, FL, College Station, TX, and San Diego, CA, southwest or southeast are better options than true south regarding to the annual energy generation. What is more, for some locations in wintertime, 90-degrees tilted PV array can generate similar or even more energy than horizontal rooftop PV arrays. These findings illustrate quantitatively the significant potential of integrating PV technology in buildings.

Keywords: Solar PV Walls · PV Electricity Generation · Building Orientation Effect · Building Integrated PV (BIPV)

1 Introduction

The application of vertical solar photovoltaic (PV) arrays is not common practice currently, but as building integrated photovoltaic (BIPV), or solar windows, technologies continue growing they can become more valuable and meaningful for buildings.

1.1 Energy Transition

It is imperative to have energy transition and more energy flexibility to face the challenge of relying heavily on fuel/oil. Recent events in the world have made this evident and worrisome. For example, in 2022 June, the gas price hit to the roof and once got as high as $7/gallon in some places of United States. Another example is the natural gas crisis still going on in Europe because countries like Germany mainly relies on Russia to import this resource. These issues demonstrated the urgency and significance for energy transition from the hydrocarbon-based energy to renewable or alternative energy sources.

D. Renné et al. (Eds.): ASES SOLAR 2023, SPE, pp. 80–86, 2023.
https://doi.org/10.1007/978-3-031-39147-7_10

1.2 Building Energy Consumption and Greenhouse Gas Emission

Buildings count for more than one third annual total energy consumption and approximately forty percent of the greenhouse gas emissions [1]. Greenhouse gas is the critical factor for global climate change. Renewable energy is able to provide supplemental electricity and help to cut greenhouse gas emission and help to mitigate climate change.

1.3 Current Solar Photovoltaic Application

Temperature and solar radiation are two main factors affecting solar photovoltaics (PV) electricity production [2]. The widespread application about roof or tilted solar PV arrays have well established research developments in depth and breadth [3–5]. However, the research about orientation influence on electricity production specifically of 90-degrees tilted solar PV is scarce. It didn't get sufficient attention since vertical tilted surface are thought not able to provide as much power as solar roofs [6]. The current integration of PV technology in buildings is addressed in 80% to the roof areas and only 20% to façades [7]. In most of the commercial buildings, façade area of building envelops could be a significant opportunity to achieve net zero carbon emission of buildings by 2050 [8]. Enlarging the current research and knowledge of vertical tilted solar PV arrays has significant meaning for the BIPV (Building Integrated PV) technology application toward a sustainable future [9, 10]. Estimating the potential impact on local electricity generation for buildings can provide guidance in their application. The objective of this study is to identify quantitatively how significant PV systems on building facades can be in comparison with the energy production of same area PV panels on horizontal rooftop. This study is limited to full unblocked surfaces, it doesn't consider any shading affects, which is known can be found on most of urban building's walls, though for some tall buildings can be a very important factor to consider in the future for electricity generation.

2 Analysis Methodology

In this study, quantitative methodology was used to conduct simulations for five United States cities: Boston, MA, College Station, TX, San Diego, CA, Miami, FL and Seattle, WA. Boston, MA (Lat: 42.37°, Long: −71.06°) located at northeast of United States. College Station, TX (Lat: 30.61°, Long: −96.34°) is in the middle south. Miami, FL (Lat: 25.77°, Long: −80.18°) is at the southeast. San Diego, CA (Lat: 32.73°, Long: −117.18°) is at the southwest and Seattle, WA (47.61°, −122.34°) is at the northwest of United States. The reason to choose these five cities is that they have different latitude/longitude and locate at very different climate zones. Solar PV arrays were simulated in the scenarios of 90-degrees tilted facing different orientations (azimuth angle - γ), to south ($\gamma = 180°$), southwest ($\gamma = 225°$), west ($\gamma = 270°$), southeast ($\gamma = 135°$) and east ($\gamma = 90°$). North, northwest and northeast were not included in this study because the cities are in the north hemisphere, and it is expected that PV array facing to these orientations will not generate significant amounts of electricity. Thirty simulations, five for rooftop solar PV array generation and twenty-five for vertical tilted solar PV Array, were conducted in this study. The result of each scenario for vertical tilted PV array was compared with the baseline: horizontal rooftop PV array -tilted zero degree.

2.1 Software and Settings

The tool used for the simulation of the solar photovoltaic system electricity generation is PVWatts [11]. This software can estimate the PV panels' electricity production either hourly or monthly for many locations. The PV system used in this study is based on a standard 19% module efficiency. The PVWatts parameters used for the simulation analysis are shown in Table 1.

Table 1. Parameters Setting for PVWatts Simulations

Module Type	Standard
PV Module Efficiency	19%
Array Type	Fixed (open rack)
Array Tilt (deg)	90
System Losses (%)	14.08
Invert Efficiency (%)	96
DC to AC Size Ratio	1.2

2.2 Results

Figure 1 presents the expected annual daily average electricity- AC System- output as the function of façade orientation for five cities. Comparing the different orientations, all the five cities have some similarities; 90-degrees tilted PV arrays facing to south, southwest and southeast generate more electricity than the ones facing to west and east in general. PV arrays facing to the south in Boston, MA, and Seattle, WA, can generate more energy than those facing to other orientations. Except in Boston and Seattle, the vertical solar PV facing to southwest, or southeast, produces more energy than facing to true south in other three cities: College Station, TX, Miami, FL and San Diego, CA.

For San Diego, CA, the annual daily average electricity generation of 90-degrees tilted PV arrays facing to southwest is the highest, followed by the arrays in Boston facing to south. Seattle is slightly less than the other four cities; most probably because it is cloudy frequently. Figure 1 clearly shows that the BIPV façade application potentially works in both north and south latitude cities. In fact, the vertical tilted PV arrays facing south in northern cities, like Boston, MA, can generate more energy than the PV arrays in southern cities, like College Station, TX and Miami, FL. Also from Fig. 1, for Boston, the vertically tilted PV arrays facing to south produce the highest annual daily average output compared to the other four orientations.

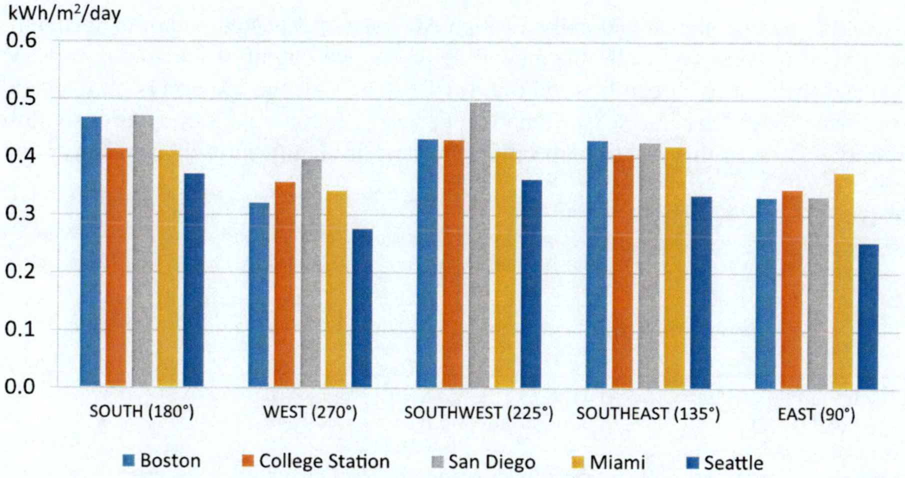

Fig. 1. Comparison of vertical PV array electricity generation for different orientations in five cities

Figure 2 includes the comparison of the monthly generation of vertical PV array facing to south with the baseline case. For Boston, vertical tilted PV array facing to south (VW-S) has advantages over horizontal rooftop PV array (HRf) in wintertime. In January, vertical tilted PV array can generate 0.65 kWh/m^2/day while rooftop PV array only generates 0.3 kWh/m^2/day. In December, vertical tilted PV can generate 0.45 kWh/m^2/day while the rooftop PV generates 0.2 kWh/m^2/day.

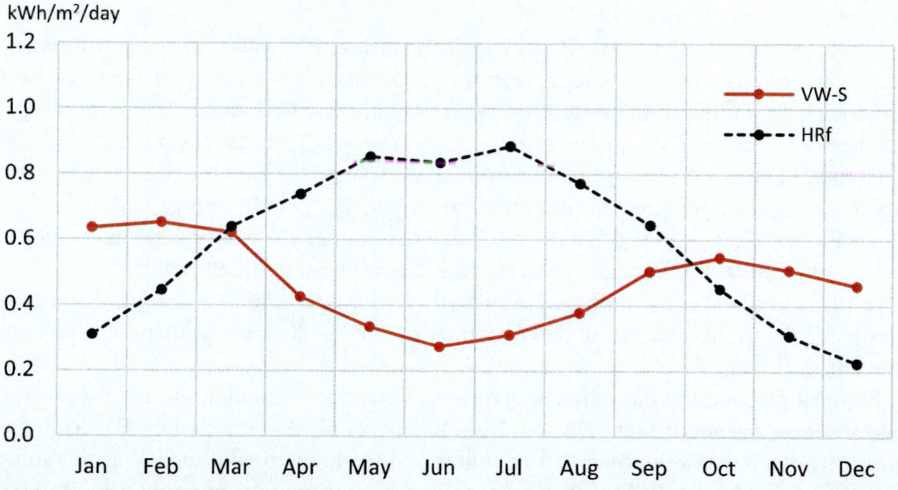

Fig. 2. Comparison of electricity generation of vertical south facing PV array and horizontal PV array in Boston, MA

Table 2 presents the annual daily average AC System expected output of vertically tilted PV array facing to south, the annual daily average output of horizontal roof PV array, and their ratios. It can be seen that for Boston, MA, the PV arrays on a façade facing to the south can achieve 80% amount of energy generation of same size of rooftop PV array. It is seen that in northern cities, the ratio is significantly higher than in the south.

Table 2. Vertical to Horizontal Annual Daily Average Electricity Generation Ratio

	Boston, MA	College Station, TX	Miami, FL	San Diego, CA	Seattle, WA
Vertical Wall - South (VW-S) [kWh/m^2/day]	0.47	0.41	0.40	0.47	0.37
Horizontal Roof (HRf) [kWh/m^2/day]	0.59	0.70	0.75	0.75	0.50
Ratio = (VW-S)/(HRf)	80%	59%	53%	63%	74%

For the environmental conditions of San Diego, CA, the vertically tilted PV array facing to southwest produces the highest annual daily average output compared to the other four orientations (see Fig. 1). The annual generation of vertically tilted PV array facing to southwest can achieve 65% of the electricity generation of same area of a horizontal rooftop PV array.

Figure 3, for San Diego, CA, includes the vertical PV wall and rooftop monthly energy generation. The red line represents the electricity generation by vertical tilted PV array facing to southwest (VW-SOUTHWEST), while the dashed black line represents the horizontal rooftop PV array electricity generation (HRf). Like Boston, it shows that in summer, the rooftop PV array has dominant advantage but in wintertime the vertical tilted PV array can generate similar amounts of energy. The difference is that for San Diego, PV wall facing to southwest maximize the annual electricity generation, but in Boston, PV wall facing to south generates the highest amount of electricity.

Furthermore, the comparative advantage of vertical PV wall over rooftop PV arrays in terms of electricity generation during winter in San Diego is not as distinct as it is for PV walls in Boston.

Figure 4 presents the monthly energy generation on five orientations and the rooftop monthly energy generation for Boston, MA. It is observed that the rooftop PV array has dominant advantage in summer. But in winter period, the vertical tilted PV array facing to southeast, southwest, and south can generate more energy than horizontal rooftop PV array.

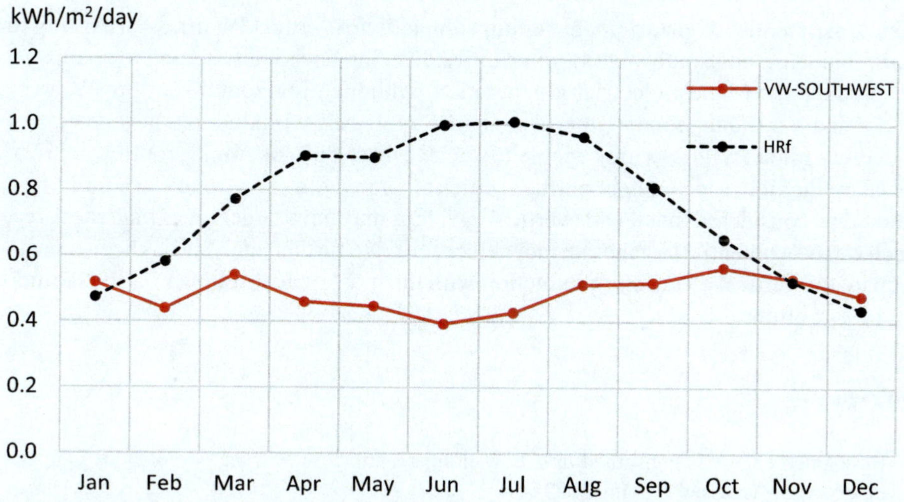

Fig. 3. Comparison of electricity generation of vertical southwest facing PV array and horizontal PV array in San Diego, CA.

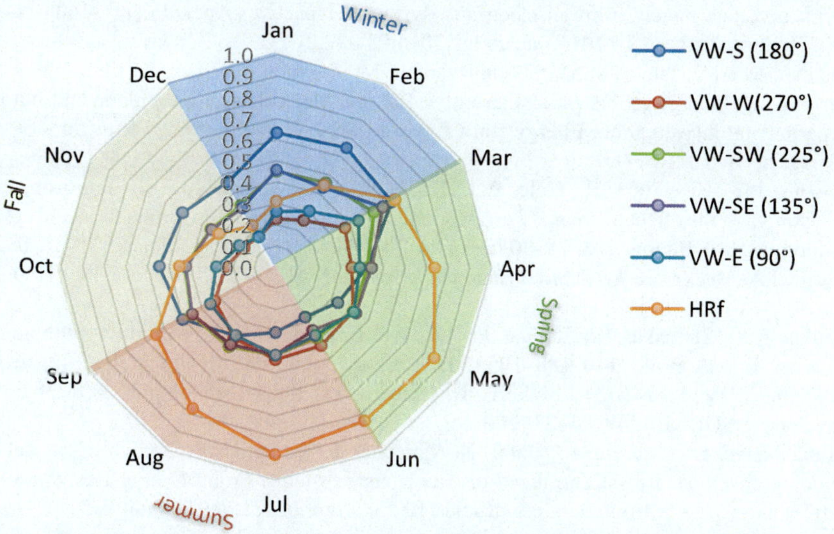

Fig. 4. Monthly average daily AC system output (kWh/m^2/day) of 90-degrees solar PV arrays facing to different orientations and horizontal roof arrays, in Boston, MA.

3 Conclusion

The simulations show that for summer periods, the vertical PV arrays in all five selected locations cannot generate as much energy as horizontal rooftop PV arrays. But in winter, the façade PV arrays are able to generate similar or even bigger amount of electricity than horizontal rooftop ones. For the five cities analyzed, the best orientations that present maximum potential of electricity generation annually for vertical PV arrays are facing to south, southeast, and southwest. In some cities, like San Diego, CA, facing to southwest will produce the highest electricity generation. Although horizontal rooftop PV array has dominant advantage in summer and annual total energy production, a same area of 90-degrees tilted PV arrays are capable to achieve up to 80% of similar rooftop area PV annual production, especially in northern cities of United States. Also in some buildings, it should be considered that façade vertical envelops may have larger areas than roof area, which can be a potential to generate more electricity. No further analysis has been done to compare the vertical electricity production with the most typical settings for the latitude tilted PV rooftop.

References

1. How much energy is consumed in U.S. buildings?. https://www.eia.gov/tools/faqs/faq.php?id=86&t=1. Accessed 15 May 2023
2. Lau, K.Y., Tan, C.W., Yatim, A.H.M.: Effects of ambient temperatures, tilt angles, and orientations on hybrid photovoltaic/diesel systems under equatorial climates. Renew. Sustain. Energy Rev. **81**, 2625–2636 (2018). https://doi.org/10.1016/j.rser.2017.06.068
3. Hartner, M., Ortner, A., Hiesl, A., Haas, R.: East to West – the optimal tilt angle and orientation of photovoltaic panels from an electricity system perspective. Appl. Energy **160**, 94–107 (2015). https://doi.org/10.1016/j.apenergy.2015.08.097
4. Mamun, M.A.A., Islam, M.M., Hasanuzzaman, M., Selvaraj, J.: Effect of tilt angle on the performance and electrical parameters of a PV module: comparative indoor and outdoor experimental investigation. Energy Built Environ. **3**(3), 278–290 (2022). https://doi.org/10.1016/j.enbenv.2021.02.001
5. Rhodes, J.D., Upshaw, C.R., Cole, W.J., Holcomb, C.L., Webber, M.E.: A multi-objective assessment of the effect of solar PV array orientation and tilt on energy production and system economics. Sol. Energy **108**, 28–40 (2014). https://doi.org/10.1016/j.solener.2014.06.032
6. Duffie, J.A., Beckman, W.A.: Solar Engineering of Thermal Process, 4th edn. Wiley, Hoboken (2013)
7. Shukla, A.K., Sudhakar, K., Baredar, P.: Recent advancement in BIPV product technologies: a review. Energy Build. **140**, 188–195 (2017). https://doi.org/10.1016/j.enbuild.2017.02.015
8. Torcellini, P.A., Pless, S.D., Judkoff, R., Crawley, D.: Solar technologies & the building envelope. ASHRAE J. **49**, 14–23 (2007)
9. Middelhauve, L., Girardin, L., Baldi, F., Maréchal, F.: Potential of photovoltaic panels on building envelopes for decentralized district energy systems. Front. Energy Res. **9**, 689781 (2021). https://www.frontiersin.org/articles/10.3389/fenrg.2021.689781/full
10. Reker, S., Schneider, J., Gerhards, C.: Integration of vertical solar power plants into a future German energy system. Smart Energy **7**, 100083 (2022). https://doi.org/10.1016/j.segy.2022.100083
11. PVWatts® Calculator. https://pvwatts.nrel.gov/pvwatts.php. Accessed 15 May 2023

The Potential of Inexpensive Solar Thermal Devices in Decarbonizing Latin American Countries' Economies

Eduardo A. Rincón-Mejía[1](✉) and Ana G. Rincón Rubio[2]

[1] Energy Program of the Autonomous University of Mexico City, 03100 Mexico City, Mexico
eduardo.rincon@uacm.edu.mx

[2] Instituto de Investigaciones Sociales de la UNAM, CU-UNAM, 04510 Mexico City, Mexico

Abstract. In Mexico, and in general, in all Latin American countries, people use fossil gas -and even firewood- to heat water for domestic use and cooking. Likewise, in most industrial applications that require low-temperature thermal processes, fossil fuels are also used. All domestic heat applications and most of the heat for processes in micro and small industries can be covered with inexpensive devices that use solar energy such as water heaters, cookers, and low-radiative flux solar concentrators commercially available. The use of these devices has a very significant potential to reduce the consumption of gas and firewood, improving the quality of the air inside dwellings and giving economic savings due to the reduction in fuel expenditure. These potentials are presented in this article with data from Mexico, with results that can be extended to most Latin American countries.

Keywords: Decarbonizing economies · inexpensive solar thermal devices · economy's decarbonization

1 The Final Consumption of Energy in Mexico

The final energy consumption in Mexico, according to the last National Energy Balance [1] published by the Mexican Ministry of Energy in 2020 was 4 383.37 PJ. Of this consumption, electric power contributed only 23.18% (1 016.07 PJ). This electric energy was generated from technologies based on fossil fuels (66.91%), hydroelectric centrals (14.64%), wind power parks (8.10%), photovoltaic installations (6.91%), nuclear power plants (1.87%), geothermal centrals (1.13%) and bioenergy units (0.44%); that is, the contribution of renewables to power generation was only 31.22%. The rest of the energy consumption was constituted by 1 813.84 PJ from fossil fuels for transportation (41.38% of the total); the heat for industrial and domestic processes added 970.48 PJ (22.14% of the total) from natural gas and LP gas, 161.31 PJ (3.68%) from petroleum and coal cokes, 20.60 PJ (0.47%) from coal, and 14.03 PJ (0.32%) from fuel oil. Biomasses contributed 370.83 PJ (8.46%, most of this in the form of firewood), and only 16.22 PJ (0.37%) was solar heat. A bleak picture, but with great opportunities to improve significantly, as mentioned below.

D. Renné et al. (Eds.): ASES SOLAR 2023, SPE, pp. 87–95, 2023.
https://doi.org/10.1007/978-3-031-39147-7_11

1.1 Thermal Consumption in Industry, Homes, and the Commercial Sectors

In Mexico, the industrial, residential, and commercial sectors present an energy consumption that is inefficient, expensive, dependent on foreign sources, and based on fossil fuels, mostly imported. According to the already cited 2020 National Energy Balance [1], half of the country's final energy consumption comes from these three sectors, and out of this half, 62% corresponds to the burning of fossil fuels for heating. It is therefore essential, in the search of environmental sustainability and social justice, decarbonizing the economy carrying out the following actions: 1) the intensive use of renewable sources for thermal and electrical services; 2) the replacement of fossil fuels used in industry for heating by solar heat and bioenergy; 3) the bioclimatic design of buildings supported by technical construction standards oriented to energy-efficiency; and 4) achieve greater energy conversion efficiencies that result in a reduction in total energy consumption by sector. These actions would reduce the emission of toxic and greenhouse gases, reduce spending on energy imports and increase the country's energy sovereignty.

The second sector with the highest energy consumption in Mexico is industry, and the third is the residential one, both only surpassed by transportation. In these two sectors, thermal uses dominate energy demand. Likewise, fossil fuels are the main source of energy, but their availability and production present a global downward trend, and their intensive use has been shown to have negative impacts not only on public health but also on the conservation of life on the planet.

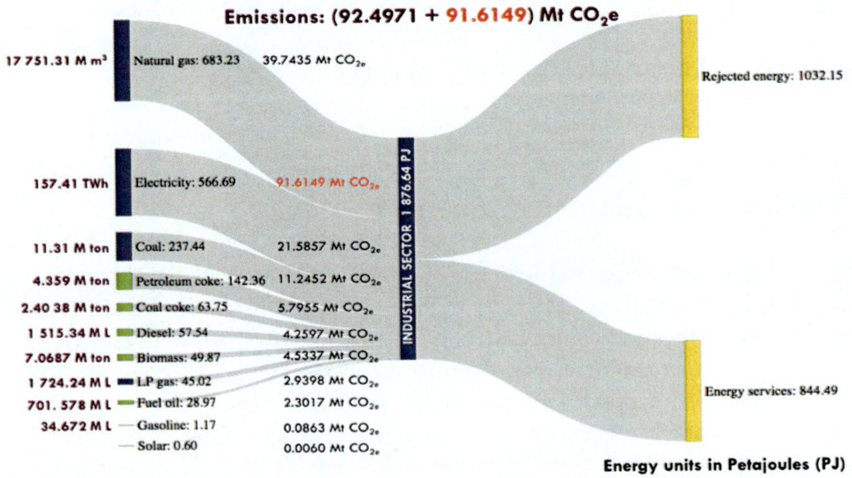

Fig. 1. Sankey diagram of the energy consumption of the industrial sector in Mexico in 2017; energy units in Petajoules. In red letters are written the indirect emissions due to electric energy consumption.

According to the data in Fig. 1, the total energy consumption of the industrial sector in 2017 was 1 877 PJ, of which: 1 251 PJ (67% of the total) correspond to fuels for industrial heat; 566.69 PJ represent electricity consumption (30.20%); and the rest (3.13%) was consumed in diesel and gasoline. Natural gas –around 70% is currently imported – contributed 54.60% of these fuels for heat in the industry. Figure 2 shows the energy consumption of the industrial sector for the previous year, 2016, according to Solar Payback 2018 [2].

Considering the emission factors of the fuels [3, 4], the direct emissions of greenhouse gases (GHG) caused by the burning of these fuels in the industrial sector amounted to 92.50 million tons of CO_{2e}, while the indirect emissions from the generation of the electrical energy consumed were 91.61 $MtCO_{2e}$ –considering that the emission factor of the Mexican electrical system during that year was 0.582 kg CO_{2e}/kWh [5]–. Thus, the equivalent CO_2 emissions of the industrial sector represented 184.11 $MtCO_{2e}$, an amount almost three times higher than the officially reported amount. The carbon intensity of this sector was 98.11 $MtCO_{2e}$/EJ.

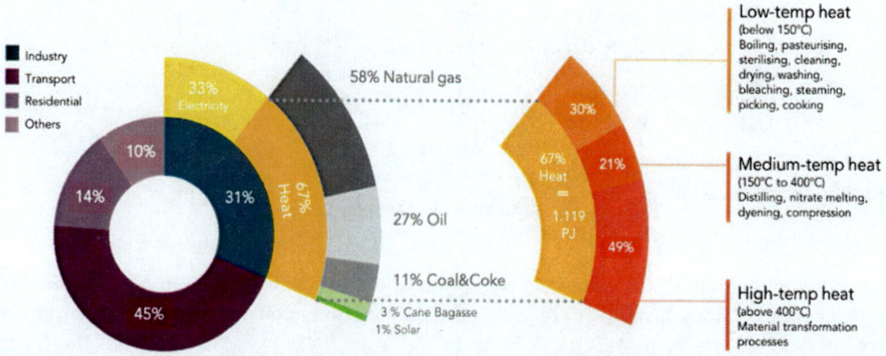

Fig. 2. Heat consumption in the industrial sector in Mexico in 2016. Source: Solar Payback 2018 [2].

Out of the total thermal consumption in the industry, 30% was heat supplied at low temperature (less than 150 °C) used in boiling, pasteurization, cleaning, drying, blanching, steaming, pickling, and cooking processes; 21% was heated at medium temperature (150–400 °C) used for distillation, coloration, and fusion of nitrates, and 49% was heat transferred at high temperature (greater than 400 °C). Considering that the adiabatic flame temperature –the case of ideal combustion without thermal losses– of the various fuels used is close to 2000 °C, from a thermodynamic point of view a lot of exergy is wasted in low and medium temperature applications, for which it would be much more efficient to use solar heat with already mature and highly profitable technologies. The use of fuels would be justifiable for higher temperature applications –in metallurgical, ceramic, cement, and glass industries–, in which biomass could be used as a complement to solar systems [6, 7], as mentioned later.

In the industrial sector, more than half of the thermal consumption is heating at low (less than 150 °C) and medium (between 150 and 400 °C) temperatures that can be supplied with solar collectors of various technologies with proven effectiveness. In this way, the consumption of fuels –fossils or biomass– can be reduced to backup use when irradiation conditions are not sufficient. A penetration of 50% of solar heat in the low and medium temperature processes would lead to an annual reduction of around 46 million tons of toxic and greenhouse effect gases.

For its part, the residential sector (see Fig. 3) consumed 751.61 PJ, distributed in almost equal parts in firewood (33.30%), LP gas (32.79%), and electricity (28.33%). The remainder (5.58%) was solar heat and natural gas consumption.

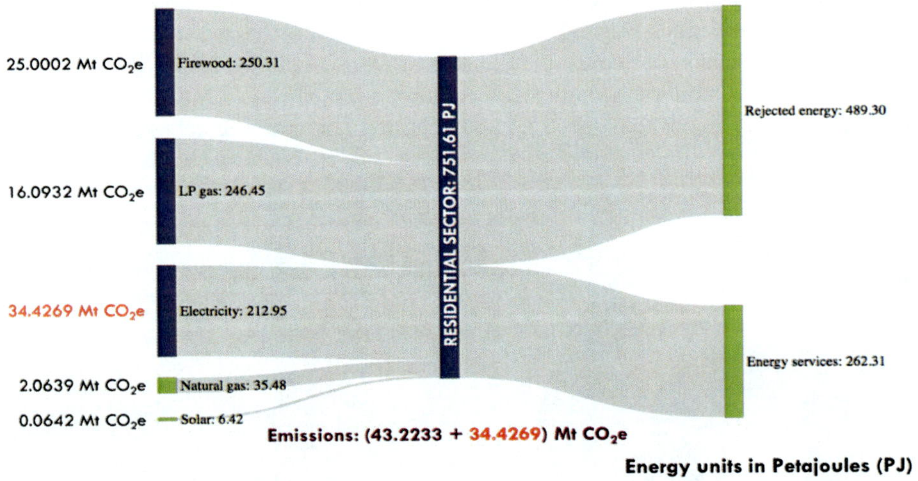

Fig. 3. Sankey diagram of the energy consumption of the residential sector in Mexico in 2017. Energy units in Petajoules, indirect emissions due to electricity consumption are written in red letters.

Direct emissions from burning fuels –essentially for heating water for bathrooms, sinks, and cooking food– was 43.22 million tons of CO_2 equivalent, which added to the 34.43 $MtCO_{2e}$ of indirect emissions in the generation of electricity consumed they threw the immense amount of 78.65 $MtCO_{2e}$. It should be noted that, because carbon dioxide is captured from the atmosphere in the photosynthetic processes of firewood formation, net emissions from its combustion are less than 26% of what is indicated in Fig. 3 by this item. All thermal applications in this sector are low temperature so the use of fuels with very high flame temperatures implies low energy efficiency; for this reason, it is in principle much healthier and more energy efficient to use solar heat for all of them, particularly for heating water for bath. However, in practice, cooking represents a much more complex energy and cultural use, since it involves numerous tasks that are carried out at different times of the day, including some in which there is not enough solar energy, such as early morning or night. For these reasons, it is recommended that, as a complement to LP gas, solar stoves be used in urban areas and that ecological wood-burning stoves be used in rural areas. The carbon intensity of the residential sector was 103.31 $MtCO_{2e}$/EJ, a figure slightly higher than that of the industrial sector.

1.2 Potential of Substituting Fossil Fuels with Solar Heat

Solar resource all over the world is several orders of magnitude greater than the entire energy consumption in all its forms, so in principle there are no limitations, in terms of the solar resource, to cover all the heat and electric demand in sunny countries like Mexico and most of Latin American countries. The average value of solar irradiation in the Mexican territory –measured on a horizontal plane– is ~5.2 kWh/m^{2}day. Stationary solar collectors are usually installed facing south and with an inclination like the latitude of the place, with which the irradiation value on them is somewhat higher than 5.5 kWh/m^{2}day in most of the territory, one of the highest values among all countries of the world.

In the industrial sector, the demand for thermal energy can be met using flat solar collectors, as well as employing parabolic trough or Fresnel-type solar concentrators which, sustainably, are technically capable of covering all the heat demand at low and medium temperatures [2]. For the rest of the thermal demand, the combustion of biomass resources could be used –e.g., agroforestry residues–, capable of reaching flame temperatures comparable to those of fossil fuels, since the high concentration thermal systems that allow these high temperatures to be reached still have costs that reduce their sustainable application niches.

For the residential sector, all the heat necessary for heating water for sanitary use can be obtained with commercial solar heaters, which are a technology with several decades of maturity: reliable, profitable, safe, and sustainable. There is no shortage in terms of finished products on the market, raw materials, or local manufacturing capacity to achieve this ideal goal, which would have many environmental, economic, social, and public health benefits. Raw materials with which water heaters for sanitary use are manufactured, both flat-plate and evacuated-tube types, are made up of the most abundant elements in the earth's crust –oxygen, silicon, aluminum, iron, etcetera–. In Mexico, as in several Latin American countries, there are various manufacturers of flat plate solar collectors – only one of them manufactures around 70 000 units each year [8]– and numerous companies import a similar number of evacuated tube systems.

In general terms, there are three technologies very available for solar water heating in the Mexican market –excluding solar concentrators, still in their early commercialization phase, as well as solar air heaters–, all of which are very efficient and profitable: 1) solar heaters with flat plate collectors, 2) solar heaters with evacuated tube collectors and 3) uncovered solar heaters for heating swimming pools. According to the most recent reports on solar heating in Mexico and the world [9], in Mexico in 2020 a little more than 1.89, 1.57, and 1.64 million square meters of collectors of these technologies were installed, respectively (5.11 million square meters in total), which added an installed thermal capacity of 3.58 thermal GW_{th}.

Various analyzes of the profitability of these systems in Mexico indicate benefit-cost ratios greater than 3 -that is, for each dollar invested there is a benefit of 3 dollars or more for gas savings-, as well as payback periods of the investment of about 2 years. In the case of industrial heat, the simple payback period for solar projects in Mexico is been reported to be between 2 and 5 years [2]. A good solar heater has a useful life of well over 10 years, so these are extremely profitable applications. It is difficult to find more profitable investments in Mexico than acquiring a solar water heater for domestic use, so it is highly recommended that each house with a roof terrace have an installation of this type. Because many families do not have enough income to invest in a solar heater, instead of subsidizing LP gas (as nowadays occurs) it would be much better -as a public policy- to look for financing schemes at zero or almost zero rate for the acquisition and installation of these systems.

Penetration of solar water heaters of barely 50% in urban dwellings – where around 80% of the highest consumption of thermal energy is for water heating – would allow savings of some 2.14 million tons of LP gas per year – given an effective calorific value of 46.16 MJ/kg for LP gas [3]–, which with an average commercial price of 1.1 USD/kg correspond to more than 2.28 billion USD per year saved to users. In addition, the annual emission of more than 6.4 million tons of carbon dioxide, nitrous dioxide, and other toxic and greenhouse gases would be avoided.

Since the adiabatic flame temperature of LP gas is around 1965 °C, using it to heat water at around 50 °C for bathing, or cooking at less than 200 °C, implies a great waste of exergy. The use of gas should be limited to high-temperature industrial applications, and only for very marginal applications in cooking food, since, as mentioned in previous paragraphs, the latter can be done with very efficient solar cookers or wood stoves, and more recently, with electric induction grills. Absurdly, nowadays in Mexico, the calorific energy of the LP gas used for cooking is more expensive than the electrical energy used for the same purpose.

Although firewood also has adiabatic flame temperatures above 1000 °C, its ease of use since immemorial time, its ability to store energy in a chemical form, and its cultural importance in cooking make it highly suitable for use in the rural environment, providing that people inside homes do not inhale the combustion fumes. This is achieved with efficient kitchen designs, which have a chimney and a well-insulated combustion chamber, which also contributes to reducing the consumption of firewood.

On the other hand, solar cookers are used by millions in countries such as China and India and can also be used massively in Mexico (and all of Latin America), given the great solar irradiation throughout the Mexican territory during most days of the year.

According to Fig. 3, the consumption of LP gas in the domestic sector in 2017 was 246.45 PJ. Out of this consumption, approximately 20% corresponded to solar cooking and the remaining 80% to water heating. Thus, the consumption of LP gas for cooking food was about 49.29 PJ. However, considering the emission factor for LP gas (about 65 299.87 $kgCO_{2e}$/TJ), cooking with LP gas causes the emission of about 3.22 million tons of equivalent carbon dioxide. If it is also considered the average calorific value of LP gas [3], in 2017 around 1 067.8 million kg of LP gas were consumed in cooking food. In this sense, a 50% penetration of solar cookers to replace half of the consumption of LP gas cookers would avoid the consumption of some 533.9 million kilograms of gas –which have a price for consumers (in May 2023) exceeding 0.42 billion USD– and the emission of around 1.61 million tons of carbon dioxide. Figure 4 shows several models of *Tolokatsin* solar ovens, with which it is possible to cook lots of food in less than 2 h on almost any day of the year [10].

To be more effective within the residential sector, actions at the level of specific technologies must be complemented with actions at the level of the dwelling. It is crucial to work on improvements at the home level based on bioclimatic design principles, since on many occasions, as in the case of thermal comfort, the only way to obtain cost-effective solutions is through improved thermal insulation, orientation, ventilation, natural lighting, and adequate housing construction materials. In particular, a very important aspect is the rescue of traditional construction techniques and materials –such as adobe, wood, palm, and others– that provide locally accessible materials with a lower energy and climate footprint than products such as concrete or steel [11].

In the commercial and services sector –whose energy consumption is an order of magnitude less than that of the industrial sector, and where half of said consumption is in the form of electricity and only 39% as heat–, the thermal demand also is carried out at low and medium temperatures, so it is feasible and sustainable to supply most of it with solar heat, using various fuels as a backup. Since the emission factor of the national electricity system nowadays is still very high –0.423 $kgCO_{2e}$/kWh [12] compared to 0.193 $kgCO_{2e}$/kWh in the United Kingdom [13]–, the recommendation to improve sustainability in shops and services is the use of on-site photovoltaic systems interconnected to the electric grid, an application that is also highly profitable and efficient.

Fig. 4. Some *Tolokatsin* solar ovens running successfully on a semi-cloudy day. Its non-image optics design allows it to concentrate part of the diffuse solar radiation to cook all kinds of food even in the rainy season.

1.3 Conclusions

To solve the challenges faced by the industrial and residential sectors, a series of technical and public policy measures must be fully promoted. Within the industrial sector, it is important to seek the gradual decarbonization of energy use by substituting fossil fuels for sources of "green heat" -solar energy and biomasses-, as well as improvements in energy efficiency that could result in a net reduction in energy consumption despite Jevons's paradox.

Likewise, it is necessary to promote the use of technologies for the use of renewable sources –photovoltaic, solar thermal, biomass, wind, and micro geothermal– in thermal and electrical uses in the residential and commercial sectors. This measure must be executed after the improvement of energy efficiency and the use of high-performance equipment subject to rigorous technical and environmental standards. Besides this, distributed energy and small and medium-sized projects should be prioritized, with the participation of local stakeholders. These actions must be accompanied in turn by the implementation –duly regulated– of thermal insulation, as well as the use of vegetation, lighting, and natural ventilation in buildings, as part of their bioclimatic design.

Given that the highest energy consumption in the form of heat in the three sectors – industrial, residential, and commercial– is carried out for low and medium-temperature applications, these could be covered efficiently, economically, and safely using solar technologies. In the residential sector, firewood is a renewable resource that has wide possibilities of use for cooking tasks and is linked to the traditions of multiple locations.

A realistic penetration of solar water heaters and solar cookers of barely 50% in the residential sector -attainable in the short term- would avoid about annually 8 million tons of carbon dioxide equivalent. An ideal total substitution of all gas and firewood boilers and stoves would increase this figure to 43.16 $MtCO_{2e}$. Supplying all low and medium temperature heat for industrial processes with solar thermal energy, the avoided emission of toxic and GHG would reach 44.07 $MtCO_{2e}$, a similar quantity to the one corresponding to the residential sector. A total (ideal) supply of all heat requirements in the commercial and services sector with solar heaters would reduce annually by 5.52 $MtCO_{2e}$.

If the recommendations described here are applied, direct GHG emissions from the industrial, residential, and commercial sectors would be significantly reduced, with many additional benefits for health, the economy, and the country's energy sovereignty and independence. It is important to highlight that the indirect emissions produced by the generation of electrical energy consumed in these sectors –which today have the same magnitude as direct emissions– can only be reduced by lowering the emissions factor of the national electrical system by increasing the participation of renewable sources in a said power generation.

As an example of public policy, it is proposed that instead of subsidizing LP gas –largely imported– for the residential sector, financing schemes be applied at zero or very reduced rates for the acquisition and installation of solar water heaters for all low-income households. This is a much more sustainable option since it would reduce the demand for gas – as well as its imports – and would significantly reduce the emission of toxic and greenhouse gases.

Even though there are substantial differences among Latin American countries in their manners for power generation for the domestic, industrial, and commercial sectors, their structures of heat consumption are very similar, so the recommendations to Mexico may be applicable to them.

References

1. Secretaría de Energía (SENER). https://www.gob.mx/sener/documentos/balance-nacional-de-energia-2020. Accessed 12 May 2023
2. Solar Payback 2018. https://www.solar-payback.com/wp-content/uploads/2018/06/Solar-Payback_Calor-Solar-para-la-Industria_Mexico.pdf. Accessed 12 May 2023
3. Instituto Nacional de Ecología y Cambio Climático (INECC). https://www.gob.mx/cms/uploads/attachment/file/110131/CGCCDBC_2014_FE_tipos_combustibles_fosiles.pdf. Accessed 10 May 2023
4. Amponsah, N.Y., Troldborg, M., Kington, B., Aalders, I., Hough, R.L.: Greenhouse gas emissions from renewable energy sources: a review of lifecycle considerations. Renew. Sustain. Energy Rev. **39**, 461–475 (2014)
5. Comisión Reguladora de Energía (CRE). https://www.gob.mx/cms/uploads/attachment/file/304573/Factor_de_Emisi_n_del_Sector_El_ctrico_Nacional_1.pdf. Accessed 12 May 2023
6. Congguang, Z., Jiaming, S., Martin, L., Ling, Q., Kang, K.: Design and simulation of a novel hybrid solar-biomass energy supply system in Northwest China. J. Clean. Prod. **233**, 1221–1239 (2019)
7. Fitsum, B.T., Ramchandra, B., Mengesha, M.: Design optimization of a hybrid solar-biomass plant to sustainably supply energy to industry: methodology and case study. Energy **220**, 119736 (2021)
8. Famerac. https://www.famerac.org. Accessed 11 May 2023
9. Weiss, W., Spörk-Dür, M.: Solar Heat Worldwide 2022. AEE Intec and IEA Solar Heating and Cooling Program, Gleisdorf, Austria (2022)
10. Rincón, E., Rincón, A., Rincón, V.: Nonimaging optics and constructal law towards optimal solar cookers. In: ASES SOLAR 2021 Proceedings. International Solar Energy Society, Boulder, CO (2021). https://doi.org/10.18086/solar.2021.01.03
11. Morillón, D., Ceballos, F.: Sustainable energy for houses, In: Rincón-Mejía, E., de las Heras, A. (eds.) Sustainable Energy Technologies, pp. 375–384. CRC Press, Boca Raton (2018)
12. Comisión Reguladora de Energía (CRE). https://www.gob.mx/cms/uploads/attachment/file/706809/aviso_fesen_2021.pdf. Accessed 14 May 2023
13. ITPenergised. https://www.itpenergised.com. Accessed 12 May 2023

Passive Freeze Protection for Thermosyphon Water Heating

Robert Stayton(✉)

Sagehill Enterprises, Santa Cruz, CA 95065, USA
bobs@sagehill.net

Abstract. Thermosyphon solar water heating systems operate with no pumps or sensors, but are vulnerable to freezing of pipes in the collectors. A previously unpublished design principle for preventing freezing has been developed and tested. It employs an additional passive circulation loop to a heat source below the collectors. As temperatures drop, the cold water in the collector sinks and is replaced by warmer water from the heat source, preventing freezing. If the heat source is a heat exchanger buried in the ground, the system is entirely passive and uses no external energy sources. This design principle will enable inexpensive thermosyphon systems to operate in a wider range of climates. Photos show a test system that has been operating for over 20 years.

Keywords: Thermosyphon · Solar DHW · Freeze Protection · Water Heating

1 Introduction

Thermosyphon water heating systems have great potential for use around the world because they are simple, reliable, inexpensive, and can be home-built. But they have always been vulnerable to the solar water heating panels freezing and bursting on cold nights, so they have been limited to consistently warm tropical climates, or they must use an inefficient and expensive heat exchanger and antifreeze combination. The design principle described here enables simple openly-plumbed thermosyphon systems to operate in temperate climates with some freezing nights, greatly expanding their potential range around the world. They can be built with simple village skills, and can operate in remote locations.

2 Classic Thermosyphon Water Heating

In a classic thermosyphon water heating system [1], a water storage tank is mounted higher than a flat plate solar collector as shown in Fig. 1.

The top of the solar collector is plumbed to the top of the storage tank, and the bottom of the collector is plumbed to the bottom of the storage tank. When the solar collector is heated by the sun, the water in the collector expands and becomes less dense. The cold water in the down pipe is therefore more dense, creating a gravitational imbalance in the

D. Renné et al. (Eds.): ASES SOLAR 2023, SPE, pp. 96–104, 2023.
https://doi.org/10.1007/978-3-031-39147-7_12

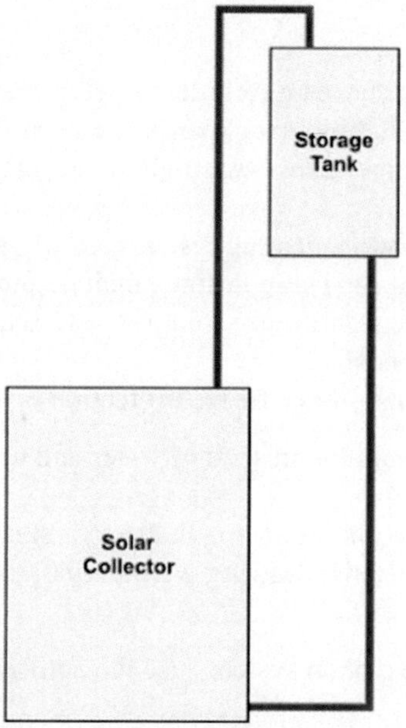

Fig. 1. Classic Thermosyphon Water Heating

plumbing loop. The cold water sinks and forces the warm water up and into the tank. As long as the coldest water in the tank is colder and more dense than the water in the collector, the water will circulate, delivering warm water to the tank.

When the collector cools off, as at night or in cloudy weather, the temperature differential is lost and the loop stops circulating, leaving the warm water at the highest point, in the storage tank. Thus, a thermosyphon water heater is self-powered by the heat of the sun, and self-regulating to turn off when the collector cools. These two features make the thermosyphon water heater a simple and reliable system.

2.1 Advantages and Disadvantages

The advantages of a classic thermosyphon system include:

- Completely automatic: water flows when the collector is hot and stops when it's not.
- No complicated parts to fail.
- No heat exchanger or toxic fluids.
- No electricity required.
- Inexpensive.
- Simple enough to do it yourself.

The disadvantages of a classic thermosyphon system include:

- The tank must be installed above the collector.
- Hard water mineral deposits can be a problem in open systems.
- Slow circulation, so less efficient than a pumped system.
- Prone to freezing.

2.2 The Freezing Problem

Classic thermosyphon systems have an Achilles' Heel: freezing. Water expands when it freezes, and the pressure created by that expansion can burst the pipes in a solar water collector exposed to cold temperatures. A single cold night is enough to damage the solar collector.

For this reason, most solar water heating systems are designed with a freeze protection system to prevent bursting pipes. These designs include pumps and sensors to control water flow, and may circulate an antifreeze solution instead of plain water. Such systems are no longer simple or inexpensive.

For classic thermosyphon systems, freeze protection options have included:

- Circulating an antifreeze solution instead of water and using a heat exchanger in the storage tank.
- Adding a dribble valve to the collector that leaks small amounts of water when temperatures approach freezing, drawing warmer water into the collector to prevent freezing.

Most commercial thermosyphon systems use the antifreeze design.

3 Dual Loop Passive Freeze Protection

A different approach to freeze protection for classic thermosyphon systems can be derived from the basic principle of thermosyphoning. That principle is that cold water sinks because it is more dense. The design principle presented here creates a path for the cold water in the collector to sink away from the solar collector.

This solution adds a second water circulation loop that connects to a heat source below the collector. The extra loop is added using simple pipe T fittings. See Fig. 2.

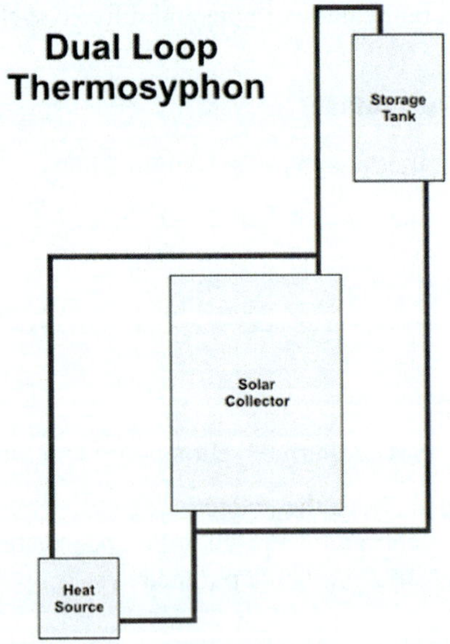

Fig. 2. Dual Loop Thermosyphon

One T fitting is added to the plumbing at the bottom of the collector, a second T fitting is added to the plumbing at the top of the collector, and direct plumbing is used between the T fittings and the heat source mounted below the collector. No check valves or pumps are used because the second loop is self-powered and self-regulating, like the upper heating loop.

This arrangement operates as follows. During the day, the upper loop is active and the system operates as a classic thermosyphon system, as shown in Fig. 3.

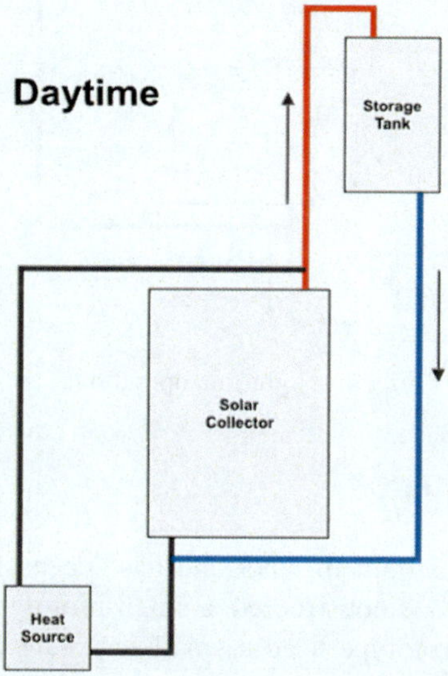

Fig. 3. Daytime operation

At night, when the temperature in the collector drops, the lower loop is activated, as shown in Fig. 4.

The second loop provides a path for the cold dense water in the collector to escape, to be replaced by warmer water from the heat source. The heat source below the collector does not have to be very hot; it just has to keep the water temperature somewhat above the maximum density of water at 4 °C (39 °F).

Providing a heat source for such mild temperatures is relatively easy. Potential sources include the following:

- Ambient heat from a building interior absorbed through a heat exchanger.
- Ground heat from a buried coil.
- Active heater with a thermostat (no longer passive, however).

The two water loops have little crossover because of the overall temperature stratification in the system. If the heat source below the collector is cooler than the water stored in the tank, then the upper loop will not be activated at night. Likewise, during the day, the lower loop will not be activated unless the cold water from the tank is colder than the lower heat source.

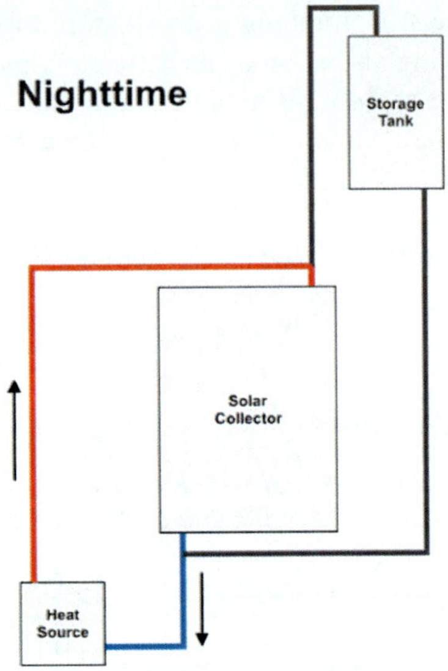

Fig. 4. Nighttime operation.

4 Demonstration System

A demonstration system was built in 1999 and has successfully operated since then. Before the full-size system was constructed, a small prototype was built to investigate the design principle. The prototype used a small flat plate collector and clear tubing instead of copper plumbing to connect to the upper storage tank and lower heat source. The heat source was a copper coil immersed in a bucket of ambient temperature water. Dye injected into the water at various points along the clear tubing showed that the two circulation loops operated as described above.

The full-size demonstration system is located in Santa Cruz in Northern California, an area that typically experiences several freezing nights per year, reaching temperatures as low as −7 °C (20 °F). The solar collectors are mounted on a hillside and are shown in Fig. 5.

The two 4 foot × 6 foot collectors are installed in parallel, and each has parallel riser tubes inside. That design provides less pipe friction than a serpentine design, an important consideration in non-pumped thermosyphon systems.

The warm water exits at the top of the collectors and is carried in a one inch copper pipe that is wrapped in insulation and buried in a trench to convey the water to the 100 gallon storage tank in the house. A second one inch copper pipe returns cold water from the bottom of the tank to the bottom of the collectors. See Fig. 6.

Fig. 5. Solar Collectors

Fig. 6. Classic thermosyphon pipes above the collectors.

The freeze-protection loop is buried in a trench below the collectors and connects to a coil of copper tubing buried in the ground, as shown in Fig. 7.

Fig. 7. Lower loop connects to buried coil.

The coil is buried 3 ft deep where the soil never reaches freezing temperatures. At night, the coil absorbs heat from the ground to keep it above freezing temperatures. When the collector water temperature drops below the ground temperature, the lower circulation loop starts up and delivers sufficient heat from the soil to the collectors to prevent them from freezing.

This system has provided a reliable supply of hot water on sunny days. During the 20+ years of operation, no pipes have ever frozen and burst, despite temperatures recorded on the surface of the collectors as low as −7 °C (20 °F). Other standing water near the collectors did freeze on such nights. The system has operated all those years with zero maintenance, other than occasionally washing the collector glass.

5 Design Considerations

Several factors should be taken into consideration when designing and building a dual loop thermosyphon system. These include:

- Climate. This design is probably not well suited to long cold winters because the heat source would be severely taxed.
- Height difference. In general, the greater the height difference between collector and heat source, the greater the flow rate, but that may mean longer pipe runs.
- Pipe diameter. The greater the pipe diameter, the greater the flow rate. Longer distances require larger pipe diameters to minimize pipe friction.
- Pipe joints and bends. As with any thermosyphon system, the system design should minimize pipe joints and bends to reduce pipe friction.
- All exposed pipes must be fully insulated.
- Sizing of the heat source.
- Reliability of the heat source.

See References [2] and [3] for more design information for classic solar thermosyphon systems.

6 Applications

This design principle for a passive freeze protection system for classic thermosyphon systems has several potential applications.

- **Do-it-yourself systems**. The implementation needs only simple plumbing skills.
- **Developing countries**. The low construction and operating costs of thermosyphon systems make them well suited for developing countries.
- **Remote locations**. The design is suitable for remote locations that lack electricity for pumping.
- **Commercial products**. Vendors of thermosyphon water heaters could replace their antifreeze fluid and heat exchanger with water, and instead install a heat source below the collectors, which are typically mounted on a flat rooftop above the living space.

7 Historical Note

Classic thermosyphon systems were popular in Southern California in the early part of the 20th century, before natural gas became available.[1] A large solar water heating industry developed with several companies providing thermosyphon systems similar to the one shown in Fig. 8, with the storage tank in the attic.

[1] This early solar history is covered thoroughly in *A Golden Thread: 2500 Years of Solar Architecture and Technology* by Ken Butti and John Perlin, Van Nostrand Reinhold; 1980.

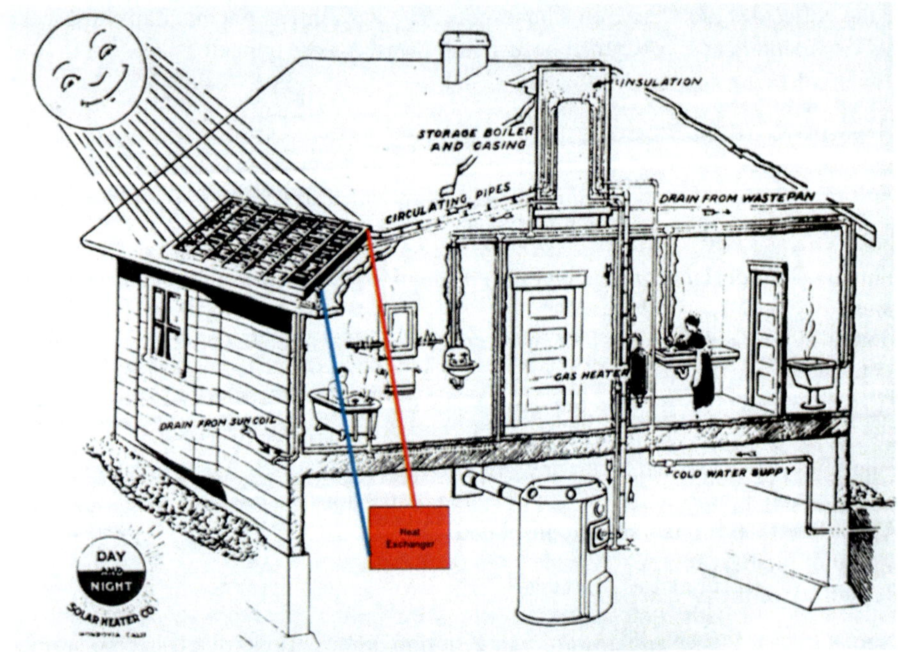

Image from A Golden Thread: 2500 Years of Solar Architecture and Technology, by Ken Butti and John Perlin

Fig. 8. Southern California thermosyphon system

These systems experienced a catastrophe on January 6, 1913, when Southern California experienced a polar vortex and saw temperatures plunge into the low 20s F. Many of the solar collectors froze and broke. The event cost the vendors a great deal of time and money to repair the systems, and damaged the reputation of solar water heating in general. If each system had had a heat exchanger heat source in their heated home mounted below the collectors as shown in the color overlay in Fig. 8, they probably would not have had frozen pipes, and some of those systems may still have been operational today.

References

1. Norton, B., Probert, S.D.: Natural-circulation solar-energy stimulated systems for heating water. Appl. Energy **11**, 167–196 (1982)
2. Colque, F.A., Siqueira, A., Silva, G.S.: A new model for sizing thermosiphon solar heating systems. Res. Soc. Dev. **9**(8), e382985673–e382985673 (2020)
3. Agbo, S.N., Unachukwu, G.O.: Design and performance features of a domestic thermosyphon solar water heater for an average-sized family in Nsukka urban. Trends Appl. Sci. Res. **2**(3), 224–230 (2007)

Investigation of Crystalline Silicon Photovoltaic Modules Degradation After 14 years Outdoor Exposure in Cologne Climate (Germany) by Electroluminescence (EL) and Infrared (IR)

I. Faye[1(✉)], A. Ndiaye[2], M. Camara[1], R Geke[3], U. Blieske[3], M. Thiame[1], and D. Kobor[1]

[1] Assane Seck University, 2700 Ziguinchor, Senegal
issafaye1211@gmail.com
[2] Cheikh Anta Diop University, Dakar, Senegal
[3] Th-Köln University, Köln, Germany

Abstract. Photovoltaic (PV) modules suffer from a variety of degradation that reduces their long-term performance and reliability. Ten polycrystalline silicon PV modules are exposed to Cologne climate. After 14 years exposure, the defects that occur at PV modules are explored by visual inspection, electroluminescence and infrared thermography. Electroluminescence (EL) characterization techniques have been exploited for the quantification of microcrack, corrosion, breakages, delamination and dark areas, while infrared (IR) has been implemented for the investigation of dark areas, hot spots and electrical losses in the PV modules. The electrical performance of the tested modules is also investigated in order to find a correlation with the visual defects.

Keywords: degradation · electroluminescence · infrared thermography · crack · PV module

1 Introduction

The analysis of data collected in the field provides information on the performance of photovoltaic modules and to understand the processes leading to the damage. PV modules are engineered to produce electricity for 25 years (Roy & Bose 2018). To ensure the long-term reliability of PV modules, quality assurance is essential to maximise financial and energy expense. The environmental factors like dust, ambient temperature, wind velocity, humidity, snowfall, hailstorms and sandstorms affect the PV modules. This implies the need for failure tracking once the modules are installed. For the reason, technical inspection methods, such as I-V curve, electroluminescence (EL) imaging and infrared (IR) thermography are applied in order to assess the quality and performance of PV modules on site (Berardone et al. 2018). EL and IR are the most widely used monitoring techniques for quality control of PV modules. EL imaging is a high-resolution characterization method that provides a map of the voltage drop from a point to another

D. Renné et al. (Eds.): ASES SOLAR 2023, SPE, pp. 105–113, 2023.
https://doi.org/10.1007/978-3-031-39147-7_13

over the solar cells (Berardone et al. 2018; Kusuma & Balakrishna 2018; Mundt et al. 2019; Sharma et al. 2019). Voltage drops are usualy caused by defects or cracks, which induce a localized additional resistance. Furthermore, EL images visually detect also intrinsic (grain boundaries, dislocations, shunt or process failures) and extrinsic physical defects (cracks, interrupted contacts, etc.) (Polizzotti 2016; Roy & Bose 2018). In the EL test, the modules operate under forward bias like a light emitting diode (LED). Excitation current can be lower or equal to I_{sc}. The injection of minority carriers to the base of the solar cell results in the emission of photons from the solar cell. These photons are detected by the camera which shows the intrinsic and extrinsic defects that decreases the minority carrier concentration. The EL technique also helped in analysing the difference in cell bias voltage in the module by observing the uneven discoloration. The aim of this study is to investigate a photovoltaic module degradation after 14 years of exposure with electrolumininescence (EL), infrared (IR) and I-V curve measurements.

2 PV System Installed in Cologne (Germany)

Ten PV modules have been investigated at TH-Köln (Cologne University of Applied Sciences, Cologne Institute of Renewable Energy) (Germany), they are installed on the roof of the solar laboratory (see Fig. 1).

Fig. 1. PV system in CIRE (Cologne)

The electrical characteristics of the multicrystalline silicon modules as indicated by the manufacturer are shown in Table 1.

Table 1. Electrical characteristics of the multicrystalline silicon modules as indicated by the manufacturer

Characteristics	Détails
Type of module	Multicrystalline
Maximum power	150 Wp (±5%)
Maximum Power Point voltage	33,4 V
Maximum Power Point current	4,49 A
Short circuit current	4,97 A
Open circuit voltage	42,2 V

3 Material and Method

All the modules were exposed to field conditions in temperate climate conditions in Cologne (Germany). I-V and P-V measurement are being measured at 1000 W/m^2 and at real operating conditions. We waited for a moment when the real irradiation was at least 1000 w/m^2 to take the measurements. The degradation data was computed from the rated performance data and the newly obtained data. The modules were characterized and analysed by EL and infrared imaging technique.

Crystalline silicon solar cell cracks (Zhou et al. 2015) can be generally classified into three categories depending on their severity:

1. Micro-cracks are solar cell cracks which do not generate inactive cell areas and consequently do not reduce the EL intensity at (Isc $\geq$ Isc of the module or cell)
2. Visible cracks are defined as cell areas that exhibit an increase in the resistance across the crack and thus appear darker in low-current EL images, but higher than the background noise of the image;
3. Complete cracks defined as cell areas that show a complete electrical separation form the active cell area, which appear dark, about equal to the background noise in the low-current. The impact of the crack orientation is relatively different on the power of the module PV. In particular, a single crack that leads to an electrical separation of a relevant part of the cell can significantly reduce power output of a PV module. The different types of cracks are:
   a. dendrite crack;
   b. several directions crack: +45° crack −45° crack; Parallel to busbar; perpendicular to busbar and cross crack (line).
   c. the thermography have being done by a camera type FLR Camera 320 * 240 pixel. in outdoors (1000 W/m^2) provide by the real irradiation

4 Results and Discussion

4.1 Thermography Imaging

The thermography imaging shows that, the modules do not have homogeneously the same temperature on the front side. Hot spots are found on the module N6; MN7 and N9. (see Fig. 2).

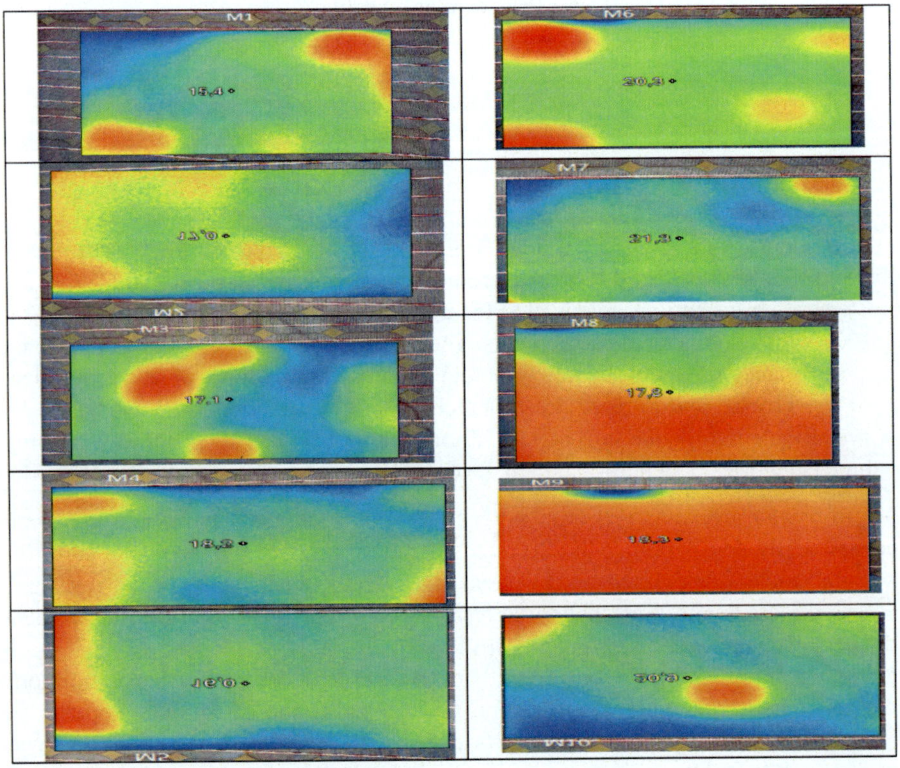

Fig. 2. Thermography of the ten multicrystalline PV module

4.2 Electroluminescence Imaging

The Table 2 shows the different defects observed with EL after 14 years exposure in moderate climate. The module M_1 and N_7 have a single crack that leads to an electrical separation of a relevant part of the circuit. These cracks lead to a moderate reduction in power of the PV module. The module M_4 have a dendritic crack that leads to high criticality power loss of the PV module (Kajari-Schröder et al. 2011) (Fig. 3).

Table 2. Electroluminescence imaging with different defects in the PV modules at Cologne

Name	Module		Module/cell
Finger interruptions	M8	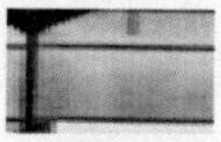	M8 (1b)
Degradation of cell interconnection	M6	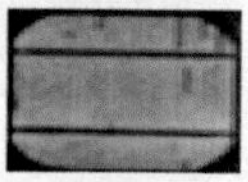	M6(2d)
Humidity corrosion in low damaged solar cell	M1 ; M2 ; M6		M1 (2f) ; M2 (8b) ; M6 (3f)
Striation ring	M2		M2 (12b)
Dendritic crack	N4		M4 (5e)
-45° crack	N1 ; N3 ; N7 ; N9		M1 (10b) ; M3 (9a) ; M7 (3e, 7e and 12e) ; M9 (3e and 12e)
+45° crack	M1; M7 ; M9		M1 (2e) ; M7 (7e) M9 (7e)

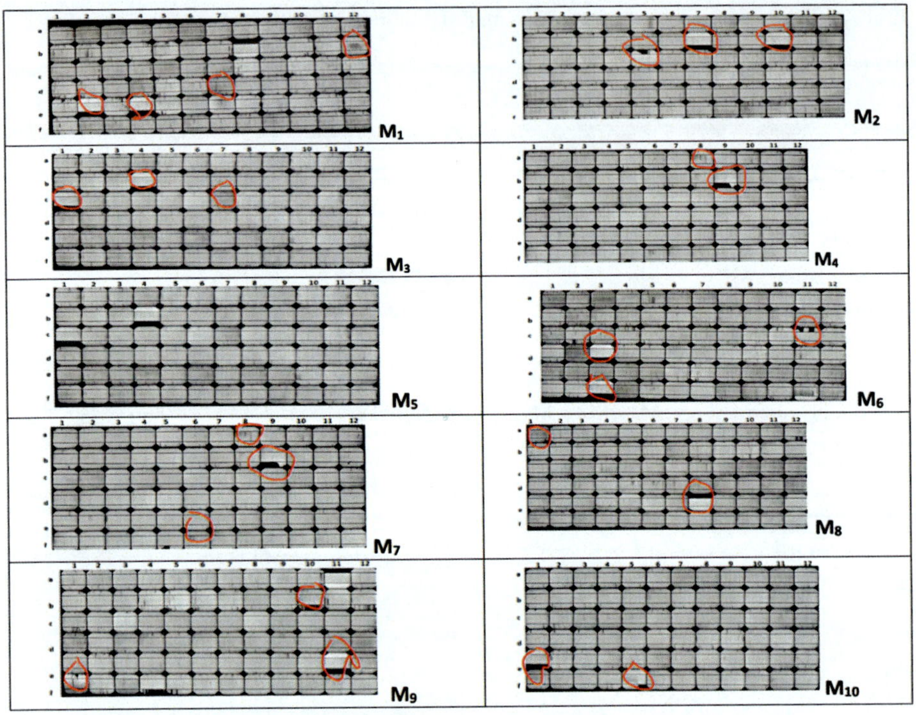

Fig. 3. Different categories of defects found in different modules investigated

4.3 Characteristics I-V and P-V

4.3.1 Degradation of Performance Parameters

The Table 3 shows the different performance parameters degradation such as, short circuit current, open circuit voltage, maximum power and fill factor. Using real irradiance in STC and dark IV the degradation of various performance parameters was found which are given in the following section.

4.3.2 Power (P_{max}) Degradation

The measurements were taken under real conditions in the temperate climate of köln where the ambient temperature was 32.37 °C.The maximal power degradation of all the module is listed in Table 3, all of them have loss of their initial maximum power. The degradation rate seems to depend on the PV module beside their same technology. Table 3 shows that, Module M4 has the highest degradation rate of 2,21% per year than the module M7 with a lower degradation of 1,85% per year. The higher degradation of the module M4 is enable to the dendritic crack observed from the electroluminescence imaging (EL) (Fig. 4).

Table 3. Degradation **of** different performance parameters

Module N°	ΔI_{sc}%	ΔV_{oc}%	ΔP_{max}%	DRP_{max}%/year	DRI_{cc}%/year	**DRFF%/year**
1	6,46	11,12	31,02	2,21	0,46	0,036
2	6,23	11,78	29,44	2,10	0,44	0,035
3	5,83	11,42	29,67	2,11	0,41	0,035
4	5,81	11,42	30,32	2,16	0,41	0,035
5	3,82	10,94	28,39	2,02	0,27	0,036
6	5,43	11,58	30,03	2,14	0,38	0,035
7	3.82	11,09	25.94	1,85	0,27	0,037
8	3,62	11,91	27,64	1,97	0,26	0,036
9	1,20	10,75	26,46	1,89	0,08	0,037
10	3,01	11,09	27,70	1,97	0,21	0,036

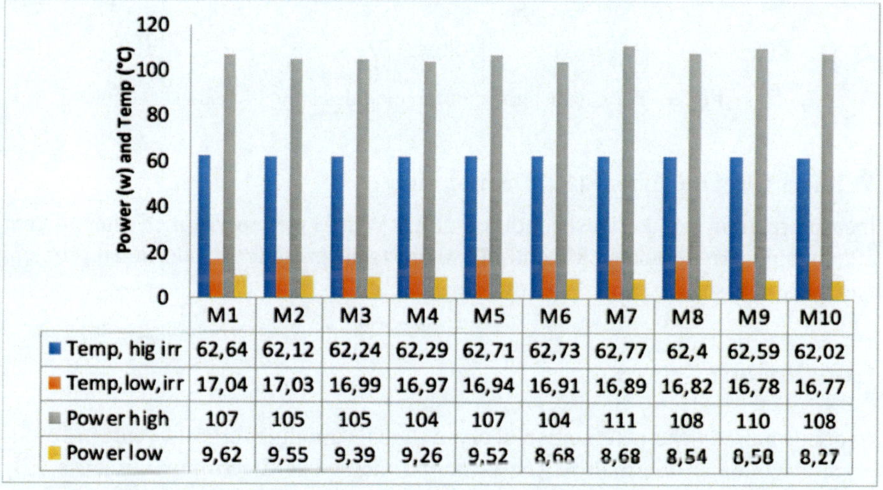

Fig. 4. Power degradation and temperature as function of the high and low irradiation

4.3.3 Fill Factor (FF) Degradation

The fill factor degradation depends relatively on the module. The lowest and highest values of annual degradation of the fill factor are 0,035 and 0,037, respectively see Table 5. The highest value of fill factor is observed on the module N7 that present hot-spot in the electrical performance analysis as presented in Fig. 5, +45°crack that electrically disconnect a part of the cell hence, decrease the short-circuit current.

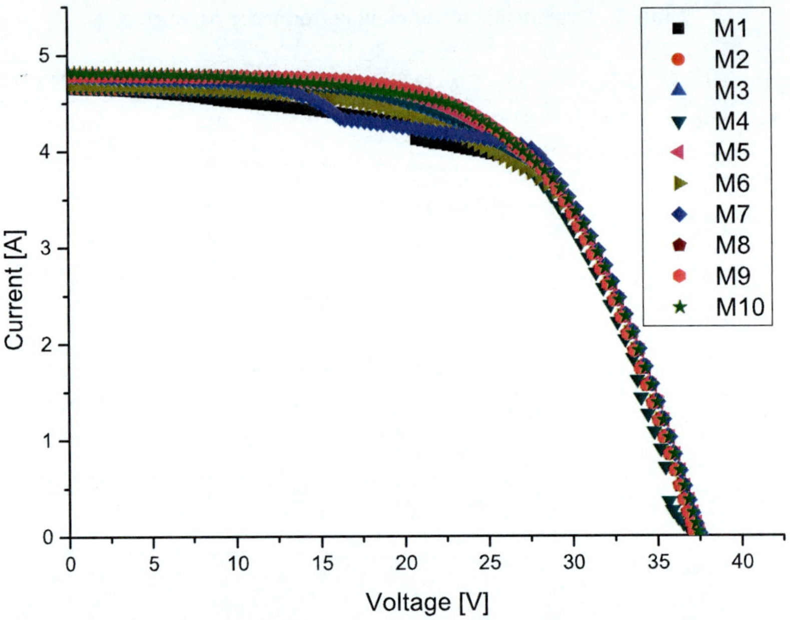

Fig. 5. I-V Characteristic of the modules after 14 years

4.3.4 Short Circuit Current (I_{sc}) Degradation

I-V measurement in standard test condition (1000 W/m^2) were performed after 14 years of outdoor exposure. Cracks in M1 and M7 suggests the presence of additional problems such as hot spots, soiling, high level of damage, etc.

5 Conclusion

This work proposes a characterization of crystalline PV modules after 14 years in outdoor exposure at Cologne (Germany). The principal method is Electroluminescence (EL), infrared imaging (IR), I-V and P-V characteristics. Results show that the maximum power degradation rate and the maximum short circuit-current degradation rate are 2,21% per year and 0,46% per year, respectively.

References

Berardone, I., Garcia, J.L., Paggi, M.: Analysis of electroluminescence and infrared thermal images of monocrystalline silicon photovoltaic modules after 20 years of outdoor use in a solar vehicle. Sol. Energy **173**, 478–486 (2018)

Köntges, M., Kurtz, S., Packard, C., Jahn, U., Berger, K.A., Kato, K.: Performance and reliability of photovoltaic systems : Subtask 3.2: Review of failures of photovoltaic modules: IEA PVPS task 13: external final report IEA-PVPS. International Energy Agency, Photovoltaic Power Systems Programme (2014)

Kusuma, J., Balakrishna, R.G.: A review on electrical characterization techniques performed to study the device performance of quantum dot sensitized solar cells. Sol. Energy **159**, 682–696 (2018)

Mundt, L.E., et al.: Quantitative local loss analysis of blade-coated perovskite solar cells. IEEE J. Photovolt. **9**(2), 452–459 (2019)

Polizzotti, A.J.: Purity matters : Enhancing carrier transport properties in tin sulfide for photovoltaic applications by reducing impurity content [PhD Thesis]. Massachusetts Institute of Technology (2016)

Roy, J.N., Bose, D.N.: Photovoltaic Science and Technology. Cambridge University Press (2018)

Sharma, J.R., Das, G., Roy, A.B., Bose, S., Mukhopadhyay, S.: Design analysis of heterojunction solar cells with aligned AZO nanorods embedded in p-type Si wafer. SILICON **12**(2), 305–316 (2019). https://doi.org/10.1007/s12633-019-00134-4

Zhou, L.M., Meng, G.W., Li, F., Wang, H.: Cell-based smoothed finite element method-virtual crack closure technique for a piezoelectric material of crack. Math. Probl. Eng. **2015**, 1 (2015). https://doi.org/10.1155/2015/371083

Kajari-Schröder, S., Kunze, I., Eitner, U., Köntges, M.: Spatial and orientational distribution of cracks in crystalline photovoltaic modules generated by mechanical load tests. Sol. Energy Mater. Sol. Cells **95**(11), 3054–3059 (2011)

Framework for Evaluation of Floating PV Project Proposals: A Guide for US Federal and State Agencies

Malvika Agarwal(✉) and Sriram Sankar Hari

Energy Science, Technology and Policy (CIT Interdisciplinary), Carnegie Mellon University, Pittsburgh, USA
malvikaa@andrew.cmu.edu

Abstract. Floating Photovoltaics (FPVs) is an upcoming extension of the solar energy sector and has been deployed in many countries to make use of the vast open water spaces available (reducing land use for PV) and to generate clean energy for nearby communities. This article outlines a framework to evaluate proposals for FPV deployment to serve communities in underdeveloped regions of the US. The state and federal governments can utilize the framework to allocate funds for floating PV projects. It considers KPIs (Key Performance Indicators), prioritizes them, and weighs each indicator to provide a definite score on a scale of 100. The KPIs cover geographic, demographic, technical, and financial metrics essential to be considered for a preliminary evaluation. It is critical for the evaluators to consider the installation's primary purpose and ensure that the proposed area is not protected or prone to natural disasters. Following that, the type of water body, annual average global hourly irradiation, area covered by PV, operating costs, and LCOE are the highest weighted metrics. Policy suggestions for FPV development in the US conclude the article as it is crucial for the state and federal governments to consider FPV-focused policies for impactful results.

Keywords: Floating photovoltaics · Community development · Evaluation framework

1 Introduction

Floating Solar Photovoltaic (FPV) technology allows for a unique opportunity for power generation by placing PV panels on bodies of still water. Unlike traditional solar panels which are installed on land or rooftops, FPV systems are designed to float on water bodies such as lakes, ponds, reservoirs, and even oceans (Beshilas n.d.).

Some of the advantages that FPV offers over traditional solutions are increased energy yield due to the cooling effect of water, reduced land area usage, and lower obstructions. FPV systems are increasingly deployed in regions with high population densities, limited availability of land, and abundant water resources. Power production gains of 1.5–22% have been documented because of the cooling effect of water on FPV

D. Renné et al. (Eds.): ASES SOLAR 2023, SPE, pp. 114–123, 2023.
https://doi.org/10.1007/978-3-031-39147-7_14

systems (Spencer et al. 2018). An FPV project typically comprises panels, floaters, combiner boxes, a mooring system, an inverter, a transformer, a cable network, a transmission system, and floating walkways ("Solar Square Blog", 2022). The mooring system is used to create a floating body of panels with power cables connected to transmission equipment on shore. FPV systems are predominantly located on static bodies of water like artificial reservoirs and dams to avoid issues arising from natural water bodies (Spencer et al. 2018). However, emerging installations are considering the viability of installation on large freshwater bodies and both off and near-shore seawater areas (Reindl and Paton 2020). FPV projects also have a faster completion time than their land-based counterparts with a 30 MW-sized project estimated to require just 8 weeks for completion (DOE 2022). The DOE prioritizes the following sectors as having the greatest potential for integration with FPV; Municipal wastewater, Agriculture, Aquaculture, and Coastal resources (DOE 2022). FPV systems reportedly have minimal impact on wildlife, except for the often-desirable reduction of algae growth (Sahu et al. 2016; Sacramento et al. 2015). FPV systems have been evaluated for potential synergies with aquaculture also known as 'Aquavoltaics' (Pringle et al. 2017).

The United States has experienced a slow adoption of FPV, but it is gaining traction in reservoir operations (hydroelectric), water treatment facilities, and residential communities (*The Economist* 2012). Although silicon PV cells have experienced an 80% cost reduction since 2009 (as predicted by Swanson's Law), the decreasing cost of deployment, particularly the cost of floaters and mooring systems has pushed for FPV development. The number of FPV projects in the US has shot up to 20 in 2020 with a total capacity of 12 MW (DC) compared to a paltry 2 projects in 2016. Yet the lack of policy incentives coupled with the abundant land area has resulted in slower growth compared to countries like Singapore which have low availability of land and ambitious renewable energy targets to meet (Ramasamy and Margolis 2021).

A detailed analysis provided by NREL quotes that 24,419 man-made water bodies in the contiguous US are suitable for FPVs (Spencer et al. 2018). Further, it was determined that covering only 27% of the area of the above-counted water bodies can produce equivalent to 10% of 2018 national generation statistics (418 TWh). The total installed capacity for FPV globally has risen from 6 megawatts (MW) to 2,579 MW from 2013 to 2021, respectively (Reindl and Paton 2020).

FPV technologies could offer various advantages, given that a significant number of these suitable bodies of water are in regions facing water scarcity, expensive land acquisition, and high electricity rates. The global FPV market is predicted to grow at a CAGR of 16.89% by 2024 and reach an estimated $775.85 million (Infiniti Research 2022). FPVs have higher upfront costs than land-based PV and hence adequate allocation of funds becomes imperative. There is an abundance of climate funds and subsidies or grants offered by state and federal agencies for solar development. Land-based PV has been explored and developed over the years, however, a holistic framework to evaluate such proposals for FPV does not exist. This paper aims to present a viable framework to evaluate the deployment of FPV solutions in a certain geographical area and assess its potential, reducing the overall complexities associated with evaluating projects in different areas with varying climate factors and economic scales and providing a consistent framework to benchmark projects. Most of these factors are common for land-based

PV as well but have been highlighted in this paper to provide a holistic approach to FPV evaluation. Additionally, these common parameters are significantly important to be considered while evaluating FPVs and hence have been mentioned.

2 Methodology

The proposals can be put forth by communities or local governments to the state or federal agencies interested in providing the funding. In most cases, such funding activity will fall under the legislation of the Department of Energy. The framework is complemented with a supporting questionnaire. The core of the model provides weights to each parameter (taken as input from the applicant or calculated) by suitable prioritization. The Excel model is a quantitative approach to deciding the worthiness of the project. Recent studies have taken only technical parameters under consideration (Spencer et al. 2018). No existing frameworks take demographic and social impact parameters into account. When considered, these parameters outweigh the financial costs for the project since the impact created by the FPV on the community can be far greater than what it costs to build the project. Hence, targeting state or federal-funded capital FPV projects in low-income communities can lead to high-impact creations and a nudge in the right direction for the future of solar in the country. The proposed floating PV must be evaluated based on certain input parameters from the applicant and calculated parameters derived from the input parameters. The authors have performed an extensive literature review and categorized the KPIs into three domains – demographic and geographic, technical, and financial. The supporting questionnaire intended for the applicants covers all the parameters. Only the key parameters have been discussed in this subsection.

2.1 Parameter Selection

Demographic and Geographic Parameters

Population and Poverty Rate. The population of the neighborhood/community intended to be receiving the benefits of the floating PV plant. The poverty rate is required to judge the current financial situation of the region. An FPV in a cumulatively low-income region may be more beneficial than a region with comparatively higher income.

Energy Consumption. Energy consumption of the region to be served highlights key points in the decision-making process. Developing a floating solar PV plant in high energy consumption region may offer greater market potential for the generated electricity. Also, higher demand can lead to better revenue prospects and potentially attract more investors. Additionally, these regions may have a more developed and robust grid system, which can facilitate plant integration. Upgrading grid infrastructure in regions with lower energy consumption may involve additional costs and logistical challenges. Some regions with lower energy consumption might have specific initiatives or incentives to promote renewable energy development. These policies can include feed-in tariffs, tax credits, or favorable regulatory frameworks that can make the project economically viable and attractive.

Type of Water Body. 7% of the country is covered by water (including coastal and inland waters) (US Geological Survey 2018). The types of water bodies which can be considered for FPV installation are preferred to be man-made since the installation and maintenance are performed in a predetermined (or controlled) environment. The following water bodies are suggested as potential sites for FPV - hydroelectric facilities, wastewater storage ponds/treatment facilities, drinking water reservoirs, quarry lakes, irrigation canals, remediation, and tailing ponds. Additionally, rivers in low-lying areas can be considered if the maximum speed of the current is lower than the threshold supported by float structures. Water bodies with high saline content are undesirable as they increase costs with the requirement of corrosion-resistant materials. The most suitable water bodies are hydroelectric facilities, irrigation canals, and wastewater treatment facilities. The collocation and operation of FPV with hydroelectric facilities have yielded multiple energy benefits, such as increased energy output, better ability to meet peak demand, and cost savings as a result of the existing transmission infrastructure (Silvério et al. 2018). On irrigation canals, FPV systems reduce water evaporation in reservoirs by reducing airflow and absorbing solar radiation that would ordinarily be absorbed by water, an attractive quality for water managers. Current solar systems on canals are not FPVs but mounted solar canopies (Ferrer-Gisbert et al. 2013; Santafé et al. 2014; Taboada et al. 2017). When used in conjunction with the facility, FPVs reduce the cost of wastewater treatment (Rosa-Clot et al. 2017; POWER 2023).

Primary Purpose. Proposals focused on community development should be given priority. Local economic development through job creation, skills development, and local procurement opportunities. Efforts should be made to prioritize local workforce participation and local supply chain engagement, enhancing socio-economic benefits.

While floating PVs serve the basic purpose of converting solar energy to electricity, FPVs can provide additional benefits. The generated electricity can be supplied directly to the grid for customer use or can serve a secondary purpose. It can be used in wastewater treatment plants to treat wastewater, directed to run electric water pumps in agricultural lands, or contribute to in-situ cleaning of polluted water bodies. The primary purpose of the proposed project by the applicant must be to serve the community in one way or another. If the project adds to the economic value of an existing professional activity, the project will receive more weightage. Existing projects in developing countries focus on in-situ water purification in polluted lakes close to areas where rainwater is the only source of replenishment for the lakes (The Times of India 2022). Sensors to monitor the physical, chemical, and biological health of the water body are placed on the floats.

Resource and Technical Parameters. The technical viability of the proposed water body to host a floating PV plant remains the first step of evaluation. Resource potential can be measured by the total area covered by the water body, the estimated area available for PV coverage and the annual average global hourly irradiation (GHI) received by the region. Other important factors pertaining specifically to the potential of the water body to sustain a viable PV structure are the water quality (saline content), snow load, wind load, wave speed, water depth, swell, and water level variation. Once the resource is examined and is considered suitable for floating PV development, technical potential parameters, namely system performance are estimated over some time. Technical potential encompasses the environmental considerations including the possible impact on the aquatic life (if any is present) of the water body and certain water bodies declared as wetlands or in protected areas. Areas of high natural disaster risk such as tsunamis, tornadoes, landslides, fires, earthquakes, droughts, floods, and heatwaves should be avoided (Lopez 2016). Certain other reliability parameters as listed below can contribute to the understanding of the evaluation committee and support decision-making through a technical perspective of the project (Teixeira 2019).

SAIFI, SAIDI, and CAIDI. SAIDI (System Average Interruption Frequency Index) is a reliability metric used in the electric power industry to measure the frequency of power outages that customers experience on average. IEEE Standard 1366-2003 defines it as "the average number of sustained interruptions per customer served during a given period". SAIDI (System Average Interruption Duration Index) is a reliability metric to measure the average duration of power outages that customers experience. IEEE Standard 1366-2003 defines it as "the average time duration of sustained interruptions per customer served during a given period". Finally, CAIDI (Customer Average Interruption Duration) is defined as "the average time duration that a customer is interrupted, calculated by dividing the total duration of all sustained interruptions by the total number of customer interruptions during a given period".

There is no centralized database that contains SAIFI, SAIDI, or CAIDI values for all utilities for different years. However, some organizations may compile this data for certain regions or countries. For example, in the United States, the Energy Information Administration (EIA) collects reliability data from utilities and publishes an annual report on electric power system reliability, which includes SAIFI, SAIDI, and CAIDI values (SAS Output, no date). Apart from the above parameters, existing grid coverage of the area and possibilities of interconnection are to be considered as well.

Economic Potential and Financial Parameters. An FPV project has economic potential when the cost required to generate the electricity is below the revenue available in terms of displaced energy and displaced capacity (Lopez 2016).

Ownership. There are 3 types of ownership. Complete private ownership is when the project is completely owned by a private entity, they have full control over the project's design, construction, and operation. They usually have less direct engagement with the local community and are perceptive as a way for private investors to profit from a public source. Complete public ownership is when government or community cooperative-owned projects allow for greater public involvement in decision-making processes and ensure that the project's objectives align with public interests. However, shortage of adequate capital, technical expertise, maintenance delays, and financial risks are a downside. Joint ownership of FPV projects allows for collaboration between public entities and private investors/partners, leveraging their respective strengths and resources. Public entities provide regulatory support and private entities contribute capital (secure financing through capital investments, loans, or PPAs) and technical expertise. In this case, shared risk mitigates financial burden.

Expected IRR. The Internal Rate of Return is an important component of project financing and developing the cash flow model. The reported values in literature so far are 6.08%,

11%, 12%, and 12.57% respectively (Sukarso and Kim 2020; Kinanti et al. 2021; Dizier, no date).

Development and Installation Cost. The cost of development and installation of a floating PV has been determined to be $1.29 per W (DC) for a 10 MW (DC) plant (Ramasamy and Margolis 2021). A rough estimate can be utilized to estimate the financial expense.

Operating Cost. Ideally, the operating costs of FPVs are expected to be greater than that for ground PV as some may require boats and divers to access floating panels away from the shore. NREL reported the levelized O&M cost as $ 15.5/kW per year calculated by considering the O&M cost for ground PV minus land leasing cost (Ramasamy and Margolis 2021).

Interconnection Cost. The cost of interconnection is an important financial parameter for the evaluation of a proposed floating PV project. Interconnection allows for connection to the grid, and net metering and enhances grid reliability.

LCOE. Inarguably one of the most important financial metrics for any energy generation project, it allows for a standardized comparison of the cost of electricity generation. Investors and project developers consider the LCOE to assess the financial viability and potential return on investment since lower LCOE values indicate more favorable economics and increased attractiveness for potential investors. LCOE is also a useful tool to optimize the financial model by varying system size, components, and operational costs.

Tax Credits and Incentives. Currently, an ITC of 30% exists on all solar projects but there are no policies specifically targeted towards floating PV development. Tax credits on floating structures and anchoring/mooring lines can heavily benefit the industry and facilitate the production of these components at lower costs.

2.2 Assignment of Weights

The Key Performance Indicators (KPIs) have been assigned weights corresponding to their perceived importance and overall impact on the project. The scale of weights is 1 to 10 where 1 is least important and 10 is most important. The assignment of weights has been done by the authors based on their understanding of the importance of the parameter after an extensive literature review. When prioritizing the KPIs based on importance for an FPV project, the primary purpose be it energy generation or water filtration takes precedence and gives an idea about the relevance and feasibility of the project. Next, factors like the available area of the project and whether the purported area is protected against development due to ecological factors need to be considered. Assessing the area for exposure to natural disasters is also a top priority. Further down the order of importance are factors like energy yield which is influenced by solar radiation measured in GHI for economic viability. Additionally, the financial metrics of the whole project including the CAPEX and OPEX need to be established first to obtain funding and determine whether the project could be economically viable or not. Since FPVs are set up in water bodies due consideration must be given to the type of water body its placed in and its impact on the aquatic ecosystem. The grid interconnection costs and complexities are affected by the proximity of the project to existing infrastructure, and metrics like SAIDI, SAIFI, and CAIDI can be used to judge the condition and requirements of the grid at a local level. Social factors are given a lower weight due to having no direct correlation with the project, however, metrics like population size can affect scale and demand, whereas the social impact of the project could have implications using the poverty rate. Finally financial parameters like type of ownership and expected Internal rate of return (IRR) are factors that while important to the balance sheet of a project, are given median weights due to their impact on the evaluation of the project and its feasibility. Multiple sources of funding and government-private partnerships are factors that could be added to this category.

The paper acknowledges that the assignment of weights and importance to factors mentioned could be arbitrarily given. However, it's important to note that the specific order of importance for the above-mentioned KPIs can vary depending on the context, project goals, and priorities. The weights and order of importance are general guidelines for an evaluation framework, but it is critical to tailor it to each specific project for its requirements and considerations. Table 1 contains a summary of the Category, KPI, Unit, Weightage, and Contribution percent of the KPI to the evaluation.

Table 1. Assigned weightage and percentage contribution for considered KPIs.

Category	KPI	Unit	Weightage	Contribution %
Demographic/Geographic	Population Size	Million	5	2.86
	Poverty Rate	%	7	4.00
	Energy Consumption	kWh	7	4.00
	Type of water body		9	5.14
	Proximity between the neighborhood center and the system	Miles	5	2.86
	Primary purpose		10	5.71
Resource/Technical	Annual average GHI	W/(m^2)	9	5.14
	Presence of aquatic life	Yes/No	7	4.00
	Protected area	Yes/No	10	5.71
	Energy Yield	kWh/kWp	8	4.57
	Area available for PV coverage	sq. Meter	9	5.14
	SAIDI, SAIFI, CAIDI		6	3.43
	Grid Coverage	% of houses with access to electricity	7	4.00
	Interconnection facility	Yes/No	8	4.57

(continued)

Table 1. (*continued*)

Category	KPI	Unit	Weightage	Contribution %
	Natural disaster prone		10	5.71
Financial	Ownership type	Public/Private/Joint	6	3.43
	Expected IRR	%	7	4.00
	Development and Installation Costs	$	8	4.57
	Operating Cost	$	9	5.14
	Interconnection Cost	$	7	4.00
	LCOE	$	9	5.14
	Total		175	100

3 Policy Suggestions

Through suitable policy development, FPVs have the potential to be an integral part of the US energy-water nexus with a focus on providing power to low-economic neighborhoods geographically close to water bodies with the potential for FPV installation. Emerging markets for FPVs like India, China, Japan, etc. to name a few have policy incentives like direct financial and production-based incentives as a measure to decrease the risk associated with FPV systems (Gadzanku 2022). Other policy metrics that could help speed up deployment and reduce barriers include water rights which is an area that requires clear-cut policies on use by FPVs and permission from the concerned bodies for their implementation. Additionally, obtaining public support through outreach programs and prioritizing public buy-in could also reduce some of the risks associated with the technology (Gadzanku 2022). Active community participation encouraging sustainable behavior will ensure clean, better-managed water bodies. Establishing community solar programs to enable residents access to affordable solar energy reduces their electricity bills and allows for active participation in the clean energy transition. FPV on polluted rivers and other water bodies can facilitate cleaning and improving water quality through 'Floating Solar Treatment' (Biomatrix 2020). Solar-powered natural water treatment systems can become a significant part of watershed restoration initiatives. Financially, apart from the existing tax credits for solar projects, financial incentives specifically tailored to support FPV projects in low-income areas including grants or low-interest loans can help offset the initial investment costs and make the project economically viable. These policy suggestions aim to address financial, social, and regulatory barriers, while also fostering community participation and benefiting low-income areas in the United States. It is crucial to customize and adapt these policies to specific local contexts and involve stakeholders in the decision-making process to ensure their effectiveness and success.

4 Limitations

Locked data with paywalls and restrictive access on the open web limited our potential to enhance the framework using market analysis for FPVs. Further, the assignment of weights is a broad generalization based on existing literature and studies, it requires tailoring to acknowledge each project's specific concerns and goals. The technical parameters mentioned are quite restricted in the public domain. Additionally, specific technical characteristics of a project could evolve over the lifetime and the framework proposed would be required to adapt to the changes.

5 Further Research

The work can be used to examine case studies on existing FPV projects and their impact on nearby communities. An extension of the existing metrics and new weights based on their spectrum can allow for better validation of different parameters. This paper can also be used for a comparative analysis between two probable locations. Additionally, the policy framework can be elaborated based on existing legislation for solar such as the Inflation Reduction Act and other granular financial parameters that can be considered during the project development phase.

Acknowledgement. The authors would like to thank Dr. Panayiotis Moutis (Formerly at Carnegie Mellon University, Pittsburgh; Currently at The City College of New York) and Dr. Barry Rawn (Carnegie Mellon University, Africa) for their continuous support and guidance during the creation of this manuscript. We would also like to thank the Energy Science, Technology, and Policy Department and Dr. Paul Salvador for funding our participation in the conference.

Declaration of Competing Interests. The authors declare no competing interests.

References

Beshilas, L.: Floating Solar Photovoltaics Could Make a Big Splash in the USA. https://www.nrel.gov/state-local-tribal/blog/posts/floating-solar-photovoltaics-could-make-a-big-splash-in-the-usa.html. Accessed 12 May 2023

Content, C.: Floating PV system provides smart energy and savings for wastewater plant. POWER Mag. (2023). https://www.powermag.com/floating-pv-system-provides-smart-energy-and-savings-for-wastewater-plant/. Accessed 16 May 2023

Department of Energy (USA) and Dawnbreaker. Market Research Study: Floating Solar Photovoltaics (2022). https://science.osti.gov/-/media/sbir/pdf/Market-Research/SETO---Floating-Solar-Photovoltaics-August-2022-Public.pdf

Dizier, A.: Techno-economic analysis of floating PV solar power plants using active cooling technique

Ferrer-Gisbert, C. et al.: A new photovoltaic floating cover system for water reservoirs. Renew. Energy **60**, 63–70 (2013). https://doi.org/10.1016/j.renene.2013.04.007

Floating solar ecosystems help to bring polluted rivers to life. Biomatrix (2020). https://www.biomatrixwater.com/news/floating-solar-ecosystems-help-to-bring-polluted-rivers-to-ife/. Accessed 16 May 23

Gadzanku, S.: Enabling Floating Solar (FPV) Deployment: Policy and Operational Considerations, United States (2022). https://www.osti.gov/servlets/purl/1875536

Global Floating Solar Panels Market 2022–2026. (n.d.). https://www.reportlinker.com/p05284859/Global-Floating-Solar-Panels-Market.html?utm_source=GNW. Accessed 15 May 2023

How Wet is Your State? The Water Area of Each State | U.S. Geological Survey (no date). https://www.usgs.gov/special-topics/water-science-school/science/how-wet-your-state-water-area-each-state#overview. Accessed 16 May 2023

Kinanti, S.P., Moeis, A.O., Kaharudin, D.: Feasibility analysis of a large scale floating photovoltaic power plant investment using financial modeling with the consideration of uncertainties factors (2021)

López, M., Soto, F., Hernández, Z.A.: Assessment of the potential of floating solar photovoltaic panels in bodies of water in mainland Spain. J. Cleaner Prod. **340**, 130752 (2022)

Pringle, A.M., Handler, R.M., Pearce, J.M.: Aquavoltaics: synergies for dual use of water area for solar photovoltaic electricity generation and aquaculture. Renew. Sustain. Energy Rev. **80**, 572–584 (2017). https://doi.org/10.1016/j.rser.2017.05.191

Ramasamy, V., Margolis, R.: Floating Photovoltaic System Cost Benchmark: Q1 2021 Installations on Artificial Water Bodies (NREL/TP-7A40-80695, 1828287, MainId:77479; p. NREL/TP-7A40-80695, 1828287, MainId:77479) (2021). https://doi.org/10.2172/1828287

Rosa-Clot, M., Tina, G.M., Nizetic, S.: Floating photovoltaic plants and wastewater basins: an Australian project. Energy Procedia **134**, 664–674 (2017). https://doi.org/10.1016/j.egypro.2017.09.585

Sacramento, E.M.D., Carvalho, P.C., de Araújo, J.C., Riffel, D.B., Corrêa, R.M.D.C., Pinheiro Neto, J.S.: Scenarios for use of floating photovoltaic plants in Brazilian reservoirs. IET Renew. Power Gener. **9**(8), 1019–1024 (2015)

Sahu, A., Yadav, N., Sudhakar, K.: Floating photovoltaic power plant: a review. Renew. Sustain. Energy Rev. **66**, 815–824 (2016). https://doi.org/10.1016/j.rser.2016.08.051

Santafé, M.R., et al.: Implementation of a photovoltaic floating cover for irrigation reservoirs. J. Clean. Prod. **66**, 568–570 (2014).: https://doi.org/10.1016/j.jclepro.2013.11.006

Silvério, N., et al.: Use of floating PV plants for coordinated operation with hydropower plants: a case study of the hydroelectric plants of the São Francisco River basin. Energy Convers. Manag. **171**, 339–349 (2018). https://doi.org/10.1016/j.enconman.2018.05.095

Spencer, R.S., Macknick, J., Aznar, A., Warren, A., Reese, M.O.: Floating photovoltaic systems: assessing the technical potential of photovoltaic systems on man-made water bodies in the continental United States. Environ. Sci. Technol. (2018). https://doi.org/10.1021/acs.est.8b04735

Sukarso, A.P., Kim, K.N.: Cooling effect on the floating solar PV: performance and economic analysis on the case of West Java Province in Indonesia. Energies **13**(9), 2126 (2020). https://doi.org/10.3390/en13092126

Reindl, T., Paton, C.: Where sun meet water: global market status, project database and economics. Keynote presented at the Solarplaza 2020 Floating Solar Conference, 3 November 2020

SETO—Floating-Solar-Photovoltaics-August-2022-Public.Pdf. Department of Energy, August 2022. https://science.osti.gov/-/media/sbir/pdf/Market-Research/SETO-Floating-Solar-Photovoltaics-August-2022-Public.pdf. Accessed 17 May 2023

Taboada, M.E., et al.: Solar water heating system and photovoltaic floating cover to reduce evaporation: experimental results and modeling. Renew. Energy **105**, 601–615 (2017). https://doi.org/10.1016/j.renene.2016.12.094

Teixeira, J.: IEEE 1366-Reliability Indices, February 2019. http://site.ieee.org/boston-pes/files/2019/03/IEEE-1366-Reliability-Indices-2-2019.pdf. Accessed May 2023

Educating and Engaging Stakeholders

Assessing Residential Solar Harvesting Potential Using the National Solar Radiation Database Across Terrestrial Elevation and Household Income

Bryan A. Kendall(✉) and Svetlana V. Poroseva

Department of Mechanical Engineering, University of New Mexico, Albuquerque, USA

Abstract. In the current study, the National Solar Radiation Database generated at the National Renewable Energy Laboratory is used to assess solar harvesting potential in Albuquerque, New Mexico. Global Horizontal Irradiance data from 2018 to 2021 is utilized in the study with the objective of investigating possible correlations between solar irradiance, median household income, and terrestrial elevation. Three zones are considered within the city's limits stretching from east to west, with each zone having an area of two square kilometers. A Python code was developed and used in the irradiance data statistical analysis. The results of the study indicate that solar irradiance values are lower at higher elevation, which is the area of highest household income. At lower elevation, solar harvesting potential is slightly higher, yet income is noticeably lower, which leads to energy justice implications associated with low-income communities having less potential to collect savings from residential solar systems within the considered region. The conclusions from the statistical analysis in combination with the yielded energy justice implications may be helpful in adjusting both renewable energy tax incentives (where applicable) and the real estate valuation process to promote a more equitable energy transition.

Keywords: Solar Harvesting Potential · Household Income · Terrestrial Elevation · Global Horizontal Irradiance · Real Estate Valuation · Equitable Energy Transition

1 Introduction

Given the increasing penetration and decreasing cost of distributed-generation photovoltaic (DGPV) systems, and the challenges associated with their variable power production, analyses of solar harvesting potential are imperative to better understand irradiance data trends as a whole and their effect on solar systems. Since market-ready DGPV systems have surfaced, which primarily consist of residential photovoltaic panels, installations have steadily grown totaling more than 500 million systems installed in the U.S. during 2021. Additionally, median installation prices fall by about $0.4/W per year over the long term in the U.S. [1]. With this solid foundation of the solar industry

D. Renné et al. (Eds.): ASES SOLAR 2023, SPE, pp. 127–135, 2023.
https://doi.org/10.1007/978-3-031-39147-7_15

in mind, methods to assess the cost and benefits of DGPV systems must evolve going forward, which foundationally includes improving data sets and the tools that use them [2].

Irradiance analysis rely on existing data sets that are primarily impacted by cloud cover prevalence. The data set used in this analysis is the National Renewable Energy Laboratory (NREL) National Solar Radiation Database (NSRDB), which is open to the public and ongoing in data acquisition [3]. The NSRDB is a solar irradiance dataset that is derived directly from the NREL Physical Solar Model [4], which in turn requires information from the National Oceanic and Atmospheric Administration (NOAA) Geostationary Operational Environmental Satellite and the National Ice Center Interactive Multisensory Snow and Ice Mapping System [5, 6]. The specific layers of the NSRDB used in this analysis and their temporal and spatial resolutions are discussed in more detail in the Methods section.

A solar harvesting potential analysis using solar irradiance data is especially relevant in New Mexico (NM) given the state has the third highest solar power potential of all states in the U.S. using sun index as the primary parameter [7, 8]. This high solar potential is compounded by high support for residential photovoltaics in Albuquerque, NM [9], making this city of interest for a regional analysis. The specific region analyzed in this study falls between the Sandia Mountains and the Rio Grande River. Albuquerque exists within a valley with the aforementioned river at its base, thus the city possesses maximum elevation differences from east to west. Three square zones within this region were considered, each two square kilometers in area and six kilometers apart. Each zone is made up of four locations (hereafter, data sampling points) that are specified using latitude and longitude. Twelve total data sampling points were considered for which solar irradiance values were extracted.

To assess potential justice implications related to solar irradiance trends, spatial median household income data can serve as a useful parameter. Income data was extracted for this study from the 2021 U.S. Census, specifically the American Community Survey 5-Year Data via an online source (censusreporter.org) [10]. In addition to income data, terrestrial elevation data was extracted from Google Earth for each data sampling point and analyzed for each zone [11].

In the current study, solar irradiance is analyzed over four years (2018–2021) to discern how it varies in Albuquerque. Irradiance values are compared with both terrestrial elevation and median household income to identify if a relationship exists between them respectively and/or in conjunction. The results of this study not only aid in developing tools that use modern data sets, but also provide important insight that can be useful to guide future DGPV system installation efforts and renewable tax incentives.

2 Methods

2.1 Zone Specification

For the purpose of conducting a solar harvesting potential analysis, three zones within Albuquerque are considered: Zone East (E), Zone Middle (M), and Zone West (W). These zones fall in districts regionally referred to as the Northeast Heights, Up Town, and the North Valley, respectively. Each zone contains four data sampling points that encompass two square kilometers. This square area was chosen given two kilometers is the highest spatial resolution offered by the NSRDB. Twelve total data sample points are used with their exact geographical locations being specified directly based on NSRDB sensor sites and therefore data availability. The region that includes all zones stretches from east to west within one percent of the latitudinal line 35.12°. The longitudinal cut off for the region is located between the Sandia Foothills and the Rio Grande River, exactly −106.49° and −106.67°. A satellite image from Google Earth with each zone superimposed is visible in Fig. 1 [11]. This area was chosen due to differences in solar ray obstruction because of the Sandia Mountains to the East, and varying elevation given Albuquerque exists within a valley that runs east to west. Both solar ray and elevation variations are indicative of varying solar irradiance values which is ideal for this analysis.

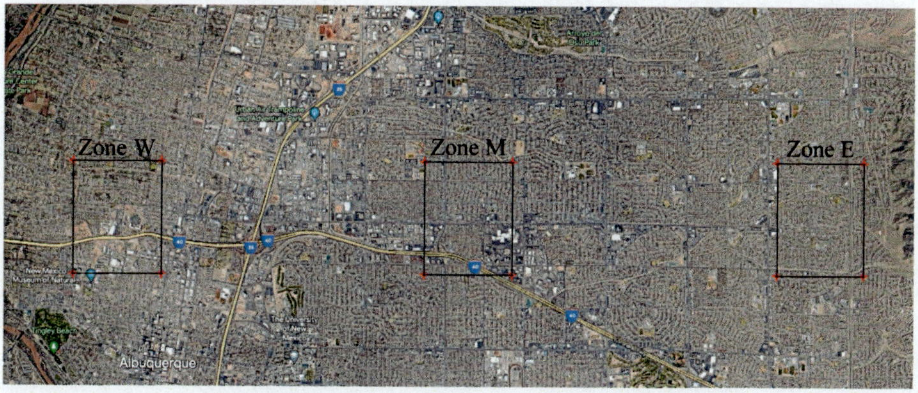

Fig. 1. Albuquerque Region with Superimposed Zones

For each data sampling point, median household income and elevation information was gathered. Income data was retrieved from an online website (censusreporter.org), which uses U.S. Census income data estimates referencing the American Community Survey 5-Year data [10]. This income data was simply compared to spatial income data present in the Energy Justice Map created by the Energy Justice Network to generally verify its accuracy [12]. Income data was extracted omitting error to represent an overall value for each data sampling point. In addition, census tract numbers were gathered for each point. Terrestrial elevation values were extracted from Google Earth for each data sampling point by hovering over their exact locations and reading values from the legend [11]. The exact location (latitude, longitude) of each data sampling point as well as which zone they fall within are presented in Table 1, alongside their relevant median household income value, census tract number, and terrestrial elevation.

2.2 Statistical Analysis

Using NSRDB data from the USA Continental and Mexico layer, a statistical analysis of Global Horizontal Irradiance (GHI) values was conducted for the 12 data sampling points visible in Table 1. This irradiance value type (GHI) includes direct and diffuse

Table 1. Data Sampling Points Information

Zone	Point #	Latitude, Longitude	Tract #	Income (USD)	Elevation (m)
Zone E	1	35.1, −106.49	1.25	$106,250	1,826
	2	35.1, −106.51	1.26	$54,107	1,728
	3	35.12, −106.49	1.07	$74,531	1,819
	4	35.12, −106.51	1.11	$107,875	1,735
Zone M	5	35.1, −106.57	6.01	$51,835	1,610
	6	35.1, −106.59	4.01	$56,050	1,592
	7	35.12, −106.57	2.06	$61,157	1,614
	8	35.12, −106.59	2.05	$51,827	1,584
Zone W	9	35.1, −106.65	27	$46,490	1,510
	10	35.1, −106.67	48	$46,298	1,510
	11	35.12, −106.65	29	$51,078	1,512
	12	35.12, −106.67	30.01	$58,929	1,511

solar radiation [13], which makes it relevant for photovoltaic and solar thermal systems. For each data sampling point, a comma separated value (CSV) file was downloaded for years 2018–2021, respectively. A total of 48 individual CSV files were downloaded, each containing annual data with an hourly temporal resolution. A Python code was developed and used in the irradiance data statistical analysis. The PANDAS Data Analysis Library within the Python programming language was employed to take averages of data over all years [14].

Data averaging took place employing two different approaches. First, yearly averages of GHI values were tabulated from Python for each data sampling point and then averaged per zone using MS Excel [15]. This first approach yielded a singular average value for each zone, for all four years analyzed, and was used to discover general irradiance trends throughout time between zones. Second, averages were taken per season within each year via Python for each data sampling point, and then tabulated in MS Excel. Seasonal averages were then averaged per zone to uncover how solar irradiance varies between zones within seasons.

A non-dimensional value (δ) was used as a metric to characterize differences in the irradiance values between zones. It was calculated as shown in Eq. (1), with x_2 representing the average seasonal GHI value of the middle or west zone, and x_1 representing the average seasonal GHI value of the east zone. Seasonal differences between zones are relevant to the current study given the variability of solar resource is an ongoing challenge in solar energy production that has the potential to destabilize electricity grids.

$$\frac{x_2 - x_1}{x_1} * 100\% = \delta \tag{1}$$

Care was taken while employing both approaches to confirm the sample size of averaged data values were consistent when taking subsequent averages. Both approaches were conducted including and excluding nighttime (zero solar irradiance) values. Excluding values at night allows averages to more precisely reflect solar irradiance in general as they are not reduced by half of the dataset being zeros.

Seasons used in the data analysis were defined using astronomical season start and end dates for the northern hemisphere. Meteorological seasons were not utilized given they vary spatially and temporally and are therefore inconsistent between years. The first day within each range that the Vernal Equinox, Summer Solstice, Autumnal Equinox, and Winter Solstice occur were used as the official start and end dates of each season. Those being, March 20th, June 20th, September 22nd, and December 21st. This span of dates to define seasons is consistent with information from the NOAA of the U.S. Department of Commerce [16], and are formally summarized in Table 2.

Table 2. Seasonal Dates Used in Statistical Analysis

Astronomical Season	Date Range
Spring	March 20th - June 19th
Summer	June 20th - September 21st
Fall	September 22nd - December 20th
Winter	December 21st - March 19th

3 Results and Discussion

Statistical averages for each three zones for all four years show that solar irradiance values are lower further east, which happens to be the area of highest median household income and terrestrial elevation (Fig. 2). This being said, the difference in GHI values is minimal between the zones (approximately 1%). Nevertheless, the negative correlation between income and irradiance data can be recognized (Fig. 2. A). The presented GHI values were calculated using the first data averaging approach which involves taking yearly averages of NSRDB data for all data sampling points, excluding nighttime values, and averaging them to yield a singular value per zone. The positive correlation between income and elevation data is demonstrated in Fig. 2. B. Income data for each zone is also averaged to create a single representational value, and is identical between both plots.

Investigation of how solar irradiance varies between zones within each season of each year yields that maximum variance exists in the summer and winter. Figure 3. Illustrates these tendencies by showing seasonal difference values for each season within each year considered. The seasonal difference values in Fig. 3 represent how solar irradiance in the west and east zones differ from each other within each season. Each difference value was calculated using the second data averaging approach, which involves NSRDB data for all data sampling points being parsed by season for each year, and then averaged excluding nighttime GHI values. These seasonal average values were then averaged per zone and used to calculate difference values between zones using Eq. (1). In Fig. 3. A. seasonal difference data are presented for the two zones, with Fig. 3. B. showing

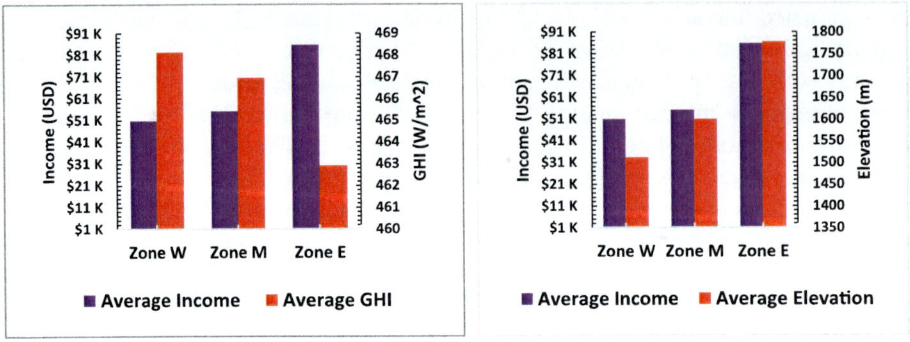

A. B.

Fig. 2. A. Average Median Household Income and Global Horizontal Irradiance, B. Average Median Household Income and Terrestrial Elevation

this data averaged over all four years. Seasonal differences between the west and east zones were found to be less than 3.5%, and did not exceed 2.5% for the middle and east zones. The highest single difference data point occurred in the summer of 2021, and the lowest difference occurred in the spring for all four years considered and with both zone pairings.

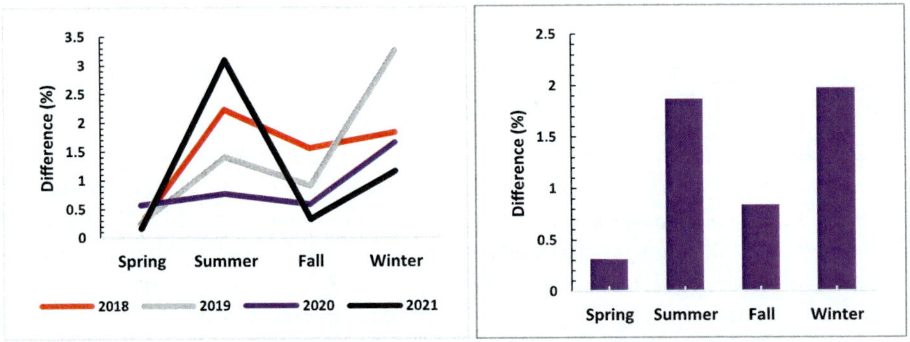

A. B.

Fig. 3. A. Seasonal Differences Between Zone West and East, B. Average Seasonal Differences Between Zone West and East

Results of the statistical analysis indicate that solar irradiance values in the zones considered increase minimally from east to west. This shows that the loss of energy incurred by solar rays traveling through more than 300 m (the maximum elevation difference between data sampling points) has a small effect on the overall irradiance trends. Low solar irradiance in Zone East could be due to solar ray obstruction during low solar angles (primarily during sunrise), as well as high occurrence of cloud coverage and weather events as a result of being in close proximity to the Sandia Mountains [17]. The higher difference in solar irradiance data between zones during the summer could be a result of a larger total span of irradiance values due to high solar intensity occurring in the summer. In the winter, high prevalence of cloud cover via storms can explain the larger solar irradiance variation. The positive correlation between income and elevation may be due to high elevation real-estate having a better view of Albuquerque, and thus being more desirable for homeowners and therefore more expensive.

This solar irradiance analysis is not the first to be conducted in NM [18, 19] or Albuquerque [20]. The topics of prior solar irradiance analyses in NM vary from assessing overarching properties of solar irradiance, to the effect of irradiation on vegetation and snow. The current study is the first that assesses income alongside solar irradiance in Albuquerque, as well as the first to use the NSRDB as its primary dataset. For this reason, the current study adds to the existing pool of literature that is related to solar irradiance in NM.

In additional to considering a limited area within a single city, other limitations of the current study are as follows. First, median household income data was collected based on the entire numbered census tract region that each specific data sampling point fell within. Average income values for each point can thus be heavily impacted by their surrounding census tract region, especially if that region is a commercial zone instead of residential. Additionally, error present in the online income source was disregarded [10]. Second, GHI being the only solar irradiance type considered has related oversights. For instance, GHI being "the sum of Direct Normal Irradiance (DNI), Diffuse Horizontal Irradiance (DHI), and ground-reflected radiation" [13], makes it not exclusively applicable to photovoltaic panels or solar thermal flat pate collectors given it includes types of irradiance that cannot be utilized by either system. Third, lacking potential for conducting an economic analysis beyond that of household income exists and became evident when attempting to amass relevant literature and find spatial data. Residential solar installation cost projections are dependent on average electricity bills that are not readily available in NM and are not represented in census data in an easily extractable format. This information is crucial in assessing the size of array needed to meet residential energy demand and therefore a limiting factor in the implementation of projected costs associated with solar panel installation and maintenance within this study. This cost is also primary in determining accurate savings potential projections that would incur after installing panels. Lastly, parameters related to environmental justice implications beyond that of median household income were not considered.

4 Conclusion

The results of the current study point towards a possible positive correlation between median household income and terrestrial elevation, and a negative correlation between median household income and solar irradiance values in Albuquerque, NM. These correlations signify that low-income residents have higher potential for solar harvesting and live at low elevations in Albuquerque. However, there is low confidence in this finding because yearly differences between zones excluding night are all approximately 1%, with seasonal differences between zones being less than 3.5%. Seasonal differences are all highest in the summer and winter, showing that solar irradiance varies the most during these seasons for all zones over all years considered. The resulting findings of the statistical analysis are preliminary in nature and require further analysis to confirm.

In the Albuquerque region discusses in this analysis as well as other areas with similar trends, implications for poor communities could be that such communities have less potential to pay for solar panels outright, and thus have less overall savings potential via the installation of residential solar systems. Lacking savings potential for low-income individuals is due to high loan/lease interest rates which hinder solar installation for those who are best positioned to harvest solar energy. Lacking savings potential can be inferred from Google Project Sunroof [21], which projects solar system installation savings based on parameters including payment plan type and average utility bills. The energy justice implications of lower income communities having both high solar harvesting potential and low savings potential regarding the installation of residential solar panels can be used to inform regional tax incentive adjustments based on solar irradiance prevalence. These potential adjustment would promote a more equitable transition to renewable energy for frontline communities.

Solar irradiance could also be used as a supplemental variable during the valuation of real-estate, alongside specification of the DGPV system that may be installed. Furthermore, the yielded result that highest irradiance differences between zones occur during the winter and summer in Albuquerque can provide insight to utility companies that have the ongoing challenge of meeting energy demand with systems that produce power variably. Potential directions for future research include additional solar irradiance/income/elevation analyses in other cities and states, power production analyses of a generalized systems to assess the effect of solar irradiance variations on simulated or real solar systems, and analyses that expand energy justice parameters to include quality of water, air, education, et cetera, to assess how they relate to solar harvesting potential. The current study thus has the potential, when combined with other analyses, to guide future work that furthers the goal of insuring a just renewable energy transition.

Acknowledgments. This material is based upon work supported in part by the National Science Foundation EPSCoR Cooperative Agreement OIA-1757207 and by the Student Training in Applied Research (STAR) internship program with the National Renewable Energy Laboratory (NREL).

References

1. Barbose, G., Darghouth, N., O'Shaughnessy, E., Forrester, S.: Tracking the sun: pricing and design trends for distributed photovoltaic systems in the United States, Lawrence Berkely National Laboratory (2022)
2. Denholm, P., et al.: Methods for analyzing the benefits and costs of distributed photovoltaic generation to the U.S. electric utility system, NREL/TP-6A20–62447, National Renewable Energy Lab. (NREL), Golden, CO, United States (2014)
3. Sengupta, M., Xie, Y., Lopez, A., Habte, A., Maclaurin, G., Shelby, J.: The national solar radiation data base (NSRDB). Renew. Sustain. Energy Rev. **89**, 51–60 (2018)
4. Physical Solar Model (PSM) v3 API | NREL: Developer Network. https://developer.nrel.gov/docs/solar/nsrdb/psm3-download/. Accessed 07 Feb 2023
5. Operations, U. D. of C. N. N. E. S. D. and I. S. O. of S. and P.: NOAA's Office of Satellite and Product Operations – GOES. https://www.ospo.noaa.gov/Operations/GOES/index.html. Accessed 07 Feb 2023

6. U.S. National Ice Center: IMS Daily Northern Hemisphere Snow and Ice Analysis at 1 Km, 4 Km, and 24 Km Resolutions, Version 1 (2004). https://nsidc.org/data/G02156/versions/1. Accessed 07 Feb 2023
7. Massachusetts Surprising Candidate for Solar Power Leadership, The Topline Strategy Group
8. Comparison of Solar Power Potential by State. https://neo.ne.gov/programs/stats/inf/201.htm. Accessed 07 Feb 2023
9. Mamkhezri, J., Thacher, J.A., Chermak, J.M.: Consumer preferences for solar energy: a choice experiment study. Energy J. **41**(5) (2020)
10. Census Reporter: Making Census Data Easy to Use. https://censusreporter.org/. Accessed 07 Feb 2023
11. Google Earth. https://earth.google.com/web/. Accessed 07 Feb 2023
12. Justice Map - Visualize Race and Income Data for Your Community. https://www.energyjustice.net/justice/index.php?gsLayer=black&gfLon=-90.245017&gfLat=38.66703601&giZoom=17&gsSize=block&giAdvanced=1. Accessed 07 Feb 2023
13. Global Horizontal Irradiance (GHI): Homer 311. https://www.homerenergy.com/products/pro/docs/3.11/global_horizontal_irradiance_ghi.html. Accessed 07 Feb 2023
14. 2023, "Python," Python.org. https://www.python.org/. Accessed 07 Feb 2023
15. 2018, "Microsoft Excel". https://www.microsoft.com/en-us/microsoft-365/excel. Accessed 07 Feb 2023
16. "Changing Seasons," NOAA Educ. https://www.noaa.gov/education/resource-collections/climate/changing-seasons. Accessed 07 Feb 2023
17. CLOUD DEVELOPMENT. https://www.weather.gov/source/zhu/ZHU_Training_Page/clouds/cloud_development/clouds.htm#CLOUD1. Accessed 07 Feb 2023
18. Gutiérrez-Jurado, H.A., Vivoni, E.R.: Ecogeomorphic expressions of an aspect-controlled semiarid basin: II. Topographic and vegetation controls on solar irradiance. Ecohydrology **6**(1), 24–37 (2013)
19. Rinehart, A.J., Vivoni, E.R., Brooks, P.D.: Effects of vegetation, albedo, and solar radiation sheltering on the distribution of snow in the Valles Caldera, New Mexico. Ecohydrology **1**(3), 253–270 (2008)
20. Garrison, J.D.: A study of solar irradiation data for six sites. Sol. Energy **32**(2), 237–249 (1984)
21. "Project Sunroof". https://sunroof.withgoogle.com/. Accessed 07 Feb 2023

Energy Conscious Construction: Curriculum Development for New Online Programs in Residential Buildings Energy-Efficiency

Omar Al-Hassawi(✉)

Washington State University School of Design and Construction, Pullman, WA 99164, USA
omar.al-hassawi@wsu.edu

Abstract. This paper reports on the curriculum development for new online educational programs at Washington State University School of Design and Construction (WSU SDC) focusing on Energy Conscious Construction (ECC) developed using funding from the U.S. Department of Energy. A review of existing programs across North America provided a snapshot of the state-of-the-art offerings in building energy efficiency. Generally, such programs are a broad survey of a wide range of topics and most of the programs are long-term post-professional master's degrees. Traditional in-person class attendance is the primary mode of delivery without many opportunities for online or distance learning. The ECC is designed to address these drawbacks and provide an interdisciplinary emphasis in energy-efficient residential buildings, covering all phases of the design process from conceptualization through construction and post-occupancy evaluation. The ECC curriculum is shaped by a series of learning modules that each focus on a specific topic. The modules are packaged into courses, and the courses are combined into three offerings: an undergraduate certificate, a graduate certificate, and a professionally oriented master's degree. The certificates address four areas of knowledge (environmental control systems, building enclosure and structural systems, simulated performance data, and performance benchmarks) whereas the master's degree addresses three additional areas of knowledge (measured performance data, smart building systems, and comprehensive design) for a total of seven areas throughout the curriculum. Two courses in the undergraduate and graduate certificates were piloted this past semester and initial student feedback has been positive with more courses being delivered this summer.

Keywords: Energy Conscious Construction · Curriculum Development · Online Education Programs · Residential Buildings · Energy Efficiency · Building Science · Zero Energy Design Designation

1 Introduction

The U.S. Department of Energy (DOE) has identified the need for workforce training that addresses five common competency gaps which exist among building professionals at a national level and are a barrier for meeting energy efficiency goals.

D. Renné et al. (Eds.): ASES SOLAR 2023, SPE, pp. 136–143, 2023.
https://doi.org/10.1007/978-3-031-39147-7_16

These are understanding building energy modeling and simulation software, meeting performance benchmarks, understanding advanced building systems, integrating smart building technologies, and using measured performance data to inform design decision making [1].

The Voiland College of Engineering and Architecture at WSU is well positioned to leverage its shared structure and in-house expertise to develop interdisciplinary programs that meet the challenges and gaps outlined above. The college has faculty experienced in all the gaps outlined above, however, current course offerings only address a subset of the gaps and are scattered across different departments and schools. Specialized courses are typically taught at the graduate level whereas undergraduate courses offer an overview of energy-efficiency topics in the built environment and are not designed with the depth that today's graduates are expected to possess.

A snapshot of the state-of-the-art educational programs related to energy-efficient buildings was established first by reviewing 50+ education programs offered by 40+ schools and colleges located in North America. The reviewed programs have a primary focus on building energy efficiency and are housed within a school that includes design-oriented programs. This helped identify the key characteristics that new programs developed at WSU SDC would need to have:

- Having a specialization: More than 95% of all reviewed programs cover a wide range of topics in energy efficient buildings and lack a focus on a specific industry sector such as residential or commercial buildings. This could potentially make it more challenging to create a common thread through the educational program and concisely communicate the learning outcomes to the students.
- Being readily accessible: There are many benefits to programs delivered in person, however, decoupling a learning opportunity from location makes it more accessible to a broader audience. Nearly 90% of the reviewed programs are delivered in person, with only five percent delivered fully online, and five percent delivered as a mix of online and in person courses. Certificates are only offered fully online, and a mixture of certificate and degree programs are offered in a hybrid mode.
- Requiring a short timeframe to complete: Most of the examined programs of study were longer term undergraduate and graduate programs. 50% of the degree programs were MS degrees and nearly a third were certificate programs predominantly at the graduate level. Minors were the primary approach to obtaining specializations at the undergraduate level but needed to be attached to a student's primary plan of study which could be a barrier for interested non-degree seeking students. While long term programs allow for in-depth inquiry on a specific topic, short term programs make it more affordable to obtain in-demand learning outcomes and create the opportunity to enter the market sooner with advanced training.

2 Proposed Programs in Energy Conscious Construction (ECC)

For the reasons above, an interdisciplinary team of faculty from the School of Design and Construction, and the School of Mechanical and Materials Engineering, and the WSU Extension Energy Program developed educational programs in Energy Conscious Construction (ECC). The ECC offers an emphasis in high-performing energy-efficient residential buildings and covers all phases of the design process from conceptualization through construction and post occupancy evaluation. The ECC uses a series of learning modules that each focus on a specific topic. The modules are packaged into courses, and the courses are combined into three offerings that are delivered fully online. These are two short-term undergraduate and graduate certificates, as well as one long-term professionally oriented master's degree.

2.1 ECC Gaps and Learning Outcomes

Based on the competency gaps identified in the funding opportunity announcement and the Pacific Northwest National Laboratories Guidelines for Building Science Education [1, 2] the seven areas of knowledge incorporated into the ECC curriculum are outlined in Table 1 below.

2.2 ECC Courses

The learning modules associated with each area of knowledge followed next with each targeting specific learning outcomes. This in turn helped identify the number of undergraduate and graduate courses required to cover all modules. A total of 12 courses were outlined, three at the undergraduate level, four spanning between the undergraduate and graduate levels, and five at the graduate level. These are broken up as follows:

1. Environmental Control Systems: Three courses were developed which are all required in the professional master's degree and the Fundamentals of Building Science course is required under all programs of study with different deliverables between undergraduate and graduate students. Environmental Control Systems I is required for the undergraduate certificate and primarily covers passive systems whereas Environmental Control Systems II is required for the graduate certificate and primarily covers active mechanical building systems.
2. Assemblies and Structures: Three courses were developed, two of which were dedicated to the undergraduate certificate and provided an overview of residential construction systems and methods. The third course (Enclosure Assemblies) is part of the graduate plans of study and provides students with in-depth working knowledge of detailing high-performing enclosures and impact on overall building performance.
3. Performance Benchmarks: One course on codes, standards, and rating systems was created and is offered in all programs of study.
4. Performance Benchmarks (Simulated): Two courses were created to cover this, one for the undergraduate level (Energy Modeling I) and another for the graduate level (Energy Modeling II), with each course tackling different software platforms.
5. Performance Benchmarks (measured): One course on field inspection was created and is only offered in the professional master's degree program.
6. Smart Technologies: One course on human-centered building design and operation was created and is only offered in the professional master's degree program.
7. Comprehensive Design: One design experience course was created and is only offered in the professional master's degree program.

Table 1. The seven areas of knowledge in the ECC curriculum.

Area of Knowledge	Gap	Outcome
1- Environmental Control Systems	Understanding the underlying fundamentals of heat, air, and moisture flows as well as incorporating passive and active systems into building designs	Obtain working knowledge in peak loads, psychometrics, and building energy balance calculations. Identify strategies of heat avoidance, heat gain, heat regulation, and heat dissipation. Analyze the impact of integrating different heat dissipation strategies on the building's overall energy consumption
2- Assemblies and Structures	Integrating enclosure assemblies aligning with the needs of the climate region. Understanding common and emerging residential buildings structural systems and their design implications	Obtain working knowledge in properly locating the layers within their designed assemblies. Understanding the environmental impacts of assembly decisions on the environment. Identify the appropriate structural system for the residential typology under consideration
3- Performance Data (Simulated)	Understanding a building's performance preconstruction and early in the design process	Obtain working knowledge in using whole-building performance simulation software. Analyze the impact of several design and user behavior factors on single- and multi-zone residential occupancies. Propose modifications to their initial designs that advance building performance
4- Performance Benchmarks	Meeting and exceeding international Energy Conservation code requirements	Understand the code creation process and the trajectory of future code modifications. Develop working knowledge in interpreting code. Modify the design to meet and exceed minimum code requirements

(*continued*)

Table 1. (*continued*)

Area of Knowledge	Gap	Outcome
5- Performance Data (Measured)	Evaluating design decisions post occupancy based on data collected on site	Develop an understanding of common field measurement equipment. Analyze measurement data for an existing residential building and propose solutions to advance performance applied onto a digital energy model of the building calibrated to utility bills
6- Smart Technologies	Understanding the fundamentals of achieving occupant visual, thermal, and acoustic comfort	Become familiar with the emerging research related to occupant comfort and impact of occupant behavior on building performance. Understand potential strategies for using both education and smart technology integration to enhance occupant behaviors, comfort, and building energy use
7- Comprehensive Design Experience	Synthesizing multiple learning outcomes into a design proposal	Develop working knowledge in creating a design of a high-performing energy-efficient residential building that meets and exceeds code requirements

2.3 Undergraduate Certificate

The undergraduate certificate is intended to be short term to fit within the students' primary programs of study, and to be completed in one semester plus a summer. It is structured to meet minimum university requirements of 15 credit hours and targets a subset of four areas of knowledge. Students complete five courses with at least one course in each area. The overall structure and the plan of study are described in Fig. 1.

2.4 Graduate Certificate

The graduate certificate is short term as well requiring one semester plus a summer to complete. It is comprised of 12 credit hours distributed over four courses and targets four areas of knowledge similar to the undergraduate certificate. The overall structure and the plan of study are described in Fig. 2.

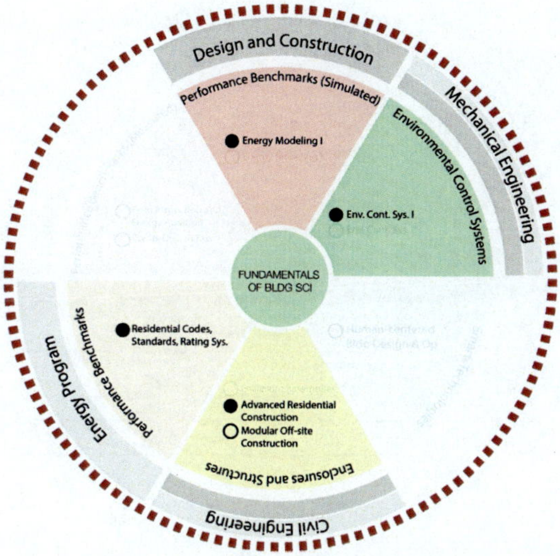

Undergraduate Certificate
15 credits

PROGRAMS
WSU
DESIGN AND CONSTRUCTION OTHER

COURSES
REQUIRED COURSES

Summer	Cr	Fall/Spring	Cr
Energy Modeling I	3	Env. Cont. Sys. I	3
Fundamentals of Building Science	3	Residential Construction OR Off-site Construction	3
Codes, Standards, Rating Systems	3		

Fig. 1. ECC undergraduate certificate overall structure and plan of study.

2.5 Professionally Oriented Master's Degree

The longest program in the ECC is the professionally oriented master's degree which is the most comprehensive program covering the three additional areas of knowledge (measured performance data, smart building technologies, and comprehensive design). It requires a minimum of one year to complete 30 credit hours distributed among ten courses. The overall structure and plan of study are described in Fig. 3.

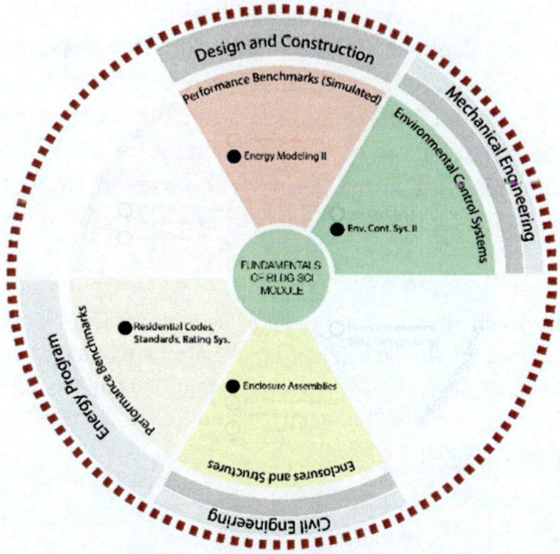

Graduate Certificate
12 credits

PROGRAMS
WSU
DESIGN AND CONSTRUCTION OTHER DISCIPLINES

COURSES
REQUIRED COURSES

Summer	Cr	Fall/Spring	Cr
Energy Modeling II	3	Enclosure Assemblies	3
Env. Cont. Sys. II	3	Fundamentals of Building Science	
Codes, Standards, Rating Systems	3		

Fig. 2. ECC graduate certificate overall structure and plan of study.

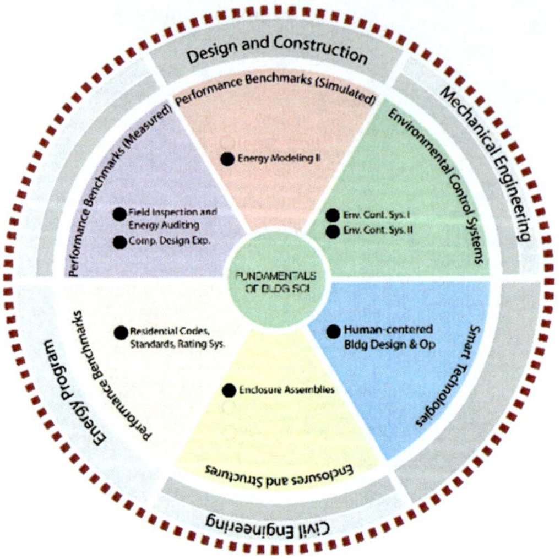

Summer	Cr	Fall	Cr	Spring	Cr
Fundamentals of Building Science	3	Energy Modeling II	3	Enclosure Assemblies	3
Codes, Standards, Rating Systems	3	Env. Cont. Sys. I	3	Env. Cont. Sys. II	3
		Human-centered Bldg Design & Op	3	Independent Study	1
		Field Inspection	4	Comprehensive Design	4

Fig. 3. ECC professionally oriented master's degree structure and plan of study.

3 Conclusions and Future Work

The first courses in the sequence have been successfully offered (Environmental Control Systems I and Enclosure Assemblies). The initial feedback received from students taking the piloted online courses has been positive. More courses need to be delivered to obtain a better understanding of the opportunities and challenges with the curriculum. Courses offered this summer are Energy Modeling I, Energy Modeling II, Fundamentals of Building Science, and Codes, Standards, and Rating Systems. Two key activities are currently being pursued. The first is aligning the curriculum with the U.S. DOE Zero Energy Design Designation requirements and exploring different iterations of assessment methods to overcome the challenge of offering students a hands-on design experience in a fully online environment [3]. And the second is establishing partnerships with other institutions so that students can bring in a percentage of the credits they have taken at their home institutions which have similar learning outcomes to the courses in the ECC certificates and have those count towards earning the certificates, thus, streamlining the path towards completion.

Acknowledgements. The author would like to deeply thank the team of faculty members working on the project. The faculty are deeply grateful for all student contributions to the project development. The faculty would also like to thank the technical advisory committee and the guest speakers who contributed to content development. Finally, the faculty would like to thank all the administrative team in charge of delivering the certificates.

References

1. U.S. Department of Energy: Funding Opportunity Announcement (FOA) Number: DE-FOA-0002196 (2021). https://eere-exchange.energy.gov/FileContent.aspx?FileID=71fa25eb-ad7c-46de-b71f-d2ce145d63b8. Accessed 12 May 2023
2. Metzger, C., Rashkin, S., Huelman, P.: Guidelines for Building Science Education. http://www.pnnl.gov/main/publications/external/technical_reports/PNNL-24143.pdf. Accessed 12 May 2023
3. U.S. Department of Energy: Zero Energy Design Designation (2023). https://www.energy.gov/eere/buildings/zero-energy-design-designation. Accessed 12 May 2023

Public Participation, Community Engagement, and Utility Regulation: Lessons and Paths Forward from Hawaii

Eric Sippert(✉)

US Department of Energy Clean Energy Innovator, Hawaii Public Utilities Commission, Honolulu, USA
eric.n.sippert.usdoe@hawaii.gov

Abstract. Public Utilities Commissions (PUCs) face increasing calls from legislatures and advocates to engage more deeply with the public through public participation and community engagement. In 2018, the Hawaii PUC identified community acceptance regarding the siting of renewable energy resources as a key barrier to reaching their 2045 100% Renewable Portfolio Standard. Without public acceptance of renewable energy projects and programs, the renewable energy transition will be slower, more difficult, and more expensive. What is—and what should be—the role of PUCs regarding community engagement and public participation? This research, first, identifies three challenges for community engagement and public participation: defining the public and community, lack of public knowledge about utility regulation and the electricity sector, and overburdened commissions with shifting needs. Second, it provides a typology of approaches to community engagement and public participation: outreach, solicitation, regulation, and alternative models. Drawing on interviews with both Hawaii and non-Hawaii government officials, intervenors, utilities and developers, and consultants, this research provides PUCs, advocates, and stakeholders with a range of options and a preliminary overview of best practices.

Keywords: Public Utilities Commission · Community Engagement · Public Participation · Renewable Energy Deployment

1 Introduction

1.1 The Changing Role of Public Utilities Commissions

Hawaii aims to reach 100% renewable energy generation by 2045. Getting to this goal will require innovative policy mechanisms and a large expansion of renewable energy generation facilities throughout the islands of Hawaii. In 2018, the Hawaii PUC identified community acceptance regarding the siting of renewable energy resources as a key barrier to reaching the Renewable Portfolio Standard [1]. Without public acceptance of renewable energy projects and programs, reaching 100% by 2045 will be difficult if not impossible. What is—and what should be—the role of the Hawaii Public Utilities Commission regarding public participation and community engagement in the transition to 100% renewable energy by 2045?

D. Renné et al. (Eds.): ASES SOLAR 2023, SPE, pp. 144–151, 2023.
https://doi.org/10.1007/978-3-031-39147-7_17

Community engagement and public participation are growing concerns for Public Utilities Commissions in general. Regulation, traditionally focused on ratemaking and project approval, now must contend with renewable portfolio standards, environmental concerns, and a grid that is becoming increasingly dynamic and decentralized. PUC proceedings usually contain legal, financial, engineering, and technical data with relevant subject experts employed by the PUC to evaluate whether proposals for new electric infrastructure, policies, and programs are in the public interest. This already considerable mandate is now joined by public and legislative calls for greater public input.

This paper is part of Department of Energy funded research that is being conducted for the Hawaii Public Utilities Commission through my role as a Clean Energy Innovator Fellow. It will culminate in a September 2023 internal report and recommendations to be presented to Commissioners and Staff. Its focus is primarily on electric utilities. This paper includes research conducted from September 2022 to April 2023.

1.2 Public Participation and Community Engagement

For the purposes of this research, *public participation* refers to interactions between the public and a PUC. This may range from formal participation in a docketed proceeding to informal participation in workshops, public comments, or other examples. *Community engagement* refers to interactions between the public and a utility or developer, which in turn may be regulated by a PUC. For example, community engagement by a developer for a new renewable energy project or by a utility for a program such as time-of-use rates.

1.3 The Hawaiian Context

The state of Hawaii is an ideal case study in many ways to study the energy transition. In terms of renewable energy, Hawaii produced 29% of its total energy production in 2022 from renewable sources with 17% of total electricity coming from solar, primarily small-scale customer-sited solar [2]. As a collection of six discrete island grids, Hawaii's two utilities (Hawaiian Electric Companies and Kauai Island Utility Cooperative) must meet each island's electrical needs without relying on electricity produced anywhere but the island where it is consumed. Furthermore, all islands are severely land constrained. New energy projects must compete with agriculture and housing for sparse available land, not to mention environmental and cultural concerns. The Hawaii PUC is a nationwide leader in innovative energy policy through Renewable Portfolio Standards (RPS), Performance-Based Ratemaking (PBF), Integrated Grid Planning (IGP), Net Metering (NEM), and Time-of-Use Rates (TOU), it has created a regulatory landscape that is often referenced from the mainland.

The combination of all these factors makes Hawaii a useful case for studying community engagement and public participation. Limited land means that new renewable energy projects are fiercely contested. New and changing programs such as NEM and TOU have the potential to confuse—and even anger—consumers. Moreover, Hawaii faces an affordability crisis that stretches into the utility sector with the average price of electricity at 41.24 cents per kilowatt hour as compared to the national average of 12.81 cents per kilowatt hour [3].

1.4 Methods

This research draws from ongoing interviews and desk research. Interviews fall into 5 categories, listed with their research purpose in Table 1 below. Desk research, which informs and is informed by interviews, has focused on identifying concrete, already-existing examples of public participation & community engagement by comparable organizations for further analysis.

Table 1. Interview Groups and Research Purpose

Interview Group	Research Purpose	Number of Interviews
Hawaii State Government Officials	Identify best practices for public participation & community engagement, understand other agencies perspective of PUC and ideal role for it	8
Non-Hawaii Public Utilities Commission Staff and Commissioners	Identify emerging trends and best practices for public participation & community engagement at other PUCs and similar organizations	6
Intervenors or Community-Based Organizations	Elucidate intervenors and community-based organizations' history and understanding of intervention at the PUC, particularly barriers	4
Utilities or Developers	Identify best practices for public participation & community engagement, understand utilities and developers' perspective of PUC and ideal role for it	3
Renewable Energy Consultant Organizations	Identify emerging trends and best practices for public participation & community engagement at other PUCs and similar organizations	4

2 Challenges

2.1 Who is the Public? Who is Community?

Traditional stakeholders in energy utility regulation include utilities, large commercial and industrial customers, energy producers, and consumer advocates. Newer actors include environmental and distributed energy advocates and activists (often energy equity or energy democracy related). The entity charged with representing the public is the consumer advocate. They face considerable challenges with their mandate of representing *all* consumers. In addition to having to represent often contradictory and competing interests, many consumer advocates are underfunded relative to other stakeholders.

Two overlapping demographics that are now more often recognized are low-income customers and disadvantaged communities. While low-income is simply an economic measure, disadvantaged communities more holistically consider environmental, economic, and health burdens, and may be expanded even more broadly to consider historical inequities. Historically, both of these demographics have been absent from PUC proceedings and considerations.

So, who is the *public* in public participation and the *community* in community engagement? Defining the community for community engagement is usually accomplished through geographic designations such as census tract, zip code, or city boundaries. Using a smaller designation may be more useful for reaching agreement and targeting benefits, but the benefits (and burdens) of new energy projects may expand much farther than a geographic designation such as census tract. How community is defined and delineated will be a site of contention for utilities, developers, advocates, communities, and PUCs as renewable energy projects continue to be built.

In an interview with a newly formed PUC Office of Public Participation, the staff member pointed to the danger of creating "new rooms" for participation "with the same faces." Ostensibly, the goal of public participation is a broader range of participants who have historically not participated in utility regulation. Reaching those groups and individuals is one challenge. Them being equipped to participate is another.

2.2 Educating Participants

PUCs, and the electric utilities they regulate, are located at the intersection of legal, policy, financial, and technical spheres. Informed participation may require knowledge of energy production, transmission, and distribution, ratemaking, RFPs, power purchase agreements (PPAs), grid planning, distributed energy, and more with many of these areas having distinct financial, technical, and social aspects. Take rooftop solar for example. Financially, PUCs and stakeholders must grapple with rooftop solar's value to the grid and the price the utility should pay for power produced from the rooftop panels among other financial considerations. Technically, the question of value to the grid is also important in addition to dealing with the complexity that decentralized energy production adds. Finally, socially, questions include how rooftop solar exacerbates or combats (energy) inequality and if poor customers without the ability to have rooftop solar are subsidizing relatively wealthy rooftop solar customers.

For any kind of useful participation, the public needs a basic understanding of how the grid works, how it is changing, and what is at stake. Enormous investor-owned utilities, NGOs, and government officials may all benefit from their expert status, but for useful public participation, education is vital. Some attempts at this education are described below in Section 3.1 Outreach.

2.3 Overburdened Commissions and Shifting Needs

As described in the introduction, the role of PUCs is changing. They are facing an increasingly complicated, quickly changing energy sector with more duties assigned to

PUCs from state legislatures. PUCs typically employ lawyers, economists, policy analysts, and engineers to analyze the legal, financial, policy, and technical implications of regulated utilities' actions. As public participation and community engagement become more important, PUCs must find staff with the skills necessary to engage in new types of evaluation and regulation.

What is the skillset necessary to foster public participation and to evaluate community engagement? PUCs need to not only communicate and educate, but also find ways to translate the public's needs and desires into the language of utility regulation. Frameworks and tools must be developed to evaluate community engagement by utilities and developers. The question of "has a community been sufficiently engaged to approve a renewable energy project?" is extremely difficult to answer.

3 A Typology of Community Engagement and Public Participation for PUCs

3.1 Outreach

PUC Outreach includes efforts by the commission to share what it does with the public, including online presence and social media, media engagement, and education. In short, this is the PUC showing the public what it does and how the public can get involved. This section describes examples of outreach at the Hawaii PUC, Vermont PUC, Public Utility Commission of Texas (PUCT), California PUC, and the Federal Energy Regulatory Commission (FERC).

As part of ongoing work to make utility regulation more accessible to the public, the Hawaii Public Utilities Commission has created pages for important dockets that attempt to explain them in a way that is accessible to a non-expert audience. Started in 2023, examples include Integrated Grid Planning, Renewable Energy Procurement, Energy Equity and Justice, and a Frequently Asked Questions page [4–7]. The Vermont PUC has gone even further with "A Citizen's Guide to the Public Utility Commission," "Public Participation and Intervention in Proceedings Before the Public Utility Commission," and "A Guide to Evidentiary Hearings."

Other PUCs have created offices specifically designed for outreach and the facilitation of public participation. The California PUC has a News and Outreach Office, which includes a Public Advisor's Office that "provides procedural information to individuals and groups who want to participate in formal proceedings" and a Business and Community Outreach Office, which focuses on outreach to stakeholders [8]. Two recently formed offices are the PUCT's Office of Public Engagement and FERC's Office of Public Participation.

Challenges for PUC-led outreach include measuring effectiveness. Statistics like webpage views provide little information and should be complemented by qualitative evidence from those who use PUC webpages. PUCs must also consider *ex parte* rules when interacting with the public. Many commissions are wary of speaking about the substance of proceedings with members of the public and violating these rules. FERC's decision to place the Office of Public Participation outside of *ex parte* rules is a step in the right direction.

3.2 Solicitation

A second potential role for engagement is solicitation by the PUC. This includes actions by the PUC that encourage stakeholders, and in particular community stakeholders, to engage with the PUC. Engagement includes formal intervention such as intervention or participation in a contested docket, rulemaking, public comments, and informal community engagement such as workshops. To encourage participation, some PUCs have used intervenor compensation [9].

Another example of solicitation of participation is the California PUC's use of oversight boards and advisory groups. The eleven-member Low Income Oversight Board was established to "advise the Commission on low-income electric and gas customer issues and to serve as a liaison for the Commission to low-income ratepayers and representatives" [10]. The Disadvantaged Communities Advisory Group reviews the California PUC and California Energy Commission's "clean energy programs and policies to ensure that disadvantaged communities, including tribal and rural communities, benefit from proposed clean energy and pollution reduction programs" [11].

Potential difficulties with solicitation efforts include budget constraints, finding representatives, and how much power to give. As the example of the California PUC's oversight boards and advisory group illustrates, commissions must consider how much to pay these groups to make it worth their time. They also must evaluate which individuals or organizations are representative of the target demographic. Finally, there is the question if boards or advisory groups will simply provide input, have veto power, or more actively shape decision-making.

3.3 Regulation

Another approach that PUCs may use is requiring utilities to lead community engagement and to then regulate that engagement. An example of this is provided by Hawaii's most recent Request for Proposals (RFP), which for the first time includes a community outreach section in its non-price criteria. Community outreach includes project description, community scoping, project benefits, government approvals, development process, community benefits package, and a communication plan. As part of the community benefits package, proposers must commit $3,000 per megawatt per year for community benefits.

A challenge of this approach is that is places utilities in the position of evaluating a developer's community engagement and then regulating it. This is new territory for the Hawaii PUC since this is the first RFP with these requirements. As selected proposals start to come before the PUC in late 2023 as PPAs, developing frameworks for evaluation will be a key challenge for the PUC as well as regulating community engagement for future RFPs.

3.4 Alternative Models

Alternative models are both the most innovative and most challenging approach to public participation and community engagement. They are the most difficult to find examples of since they are most likely to require a legislative mandate or changes in the law. PUCs are risk-averse institutions and alternative models require bold action.

A common example of an alternative model cited by energy activists is third party oversight as a form of community engagement. While different in substantial ways, ongoing attempts at creating community water authorities in Maui present several questions for renewable energy development in Hawaii. Could the PUC create island-level community boards that help evaluate renewable energy projects? Could other bodies be created that assist with bidirectional communication between the PUC and communities? Could the role of the Consumer Advocate be expanded? Another example of a potential alternative model is the Molokai Community Energy Resiliency Action Plan (CERAP), which is hosted by the Molokai Clean Energy Hui. The CERAP process, which has taken place for over a year, could provide valuable lessons for the PUC regarding community engagement. The PUC could interface with entities like Molokai Clean Energy Hui regarding renewable energy projects and planning.

4 Conclusion

Public participation and community engagement are relatively new challenges for PUCs and the energy transition more broadly. As more states implement Renewable Portfolio Standards and get closer to 100% renewable energy, public acceptance and buy-in will become increasingly important. In an interview with Hawaii-based environmental advocate, the advocate said that Hawaii's first ten years of energy transition were about "low-hanging fruit" while the second ten years will be about community engagement. PUCs cannot afford to neglect the social aspects of the energy transition. Public participation and community engagement are vital not only for a just transition, but for any transition at all.

Three challenges for PUCs are defining the public and community, a lack of knowledge about energy, and a changing mandate. This research provides a typology to classify existing public participation and community engagement, which includes outreach, solicitation, regulation, and alternative models. PUCs will most likely use a combination of these four categories moving forward. Future research may examine the effectiveness of these different types of public engagement, the procedural effects of it, and which stakeholders it privileges. One thing is certain: as the energy transition proceeds, the questions of how to engage the broader public in utility regulation is here to stay.

Acknowledgements. This research was supported in part by the Department of Energy (DOE) Office of Energy Efficiency and Renewable Energy administered by the Oak Ridge Institute for Science and Education (ORISE) for the DOE. ORISE is managed by Oak Ridge Associated Universities (ORAU) under DOE contract number DE-SC0014664. All opinions expressed in this paper are the author's and do not necessarily reflect the policies and views of DOE, ORAU, ORISE, or the Hawaii Public Utilities Commission.

References

1. State of Hawaii Public Utilities Commission, Report to the 2019 Legislature on Hawaii's Renewable Portfolio Standards (2018)

2. U.S. Energy Information Administration, Hawaii State Profile. https://www.eia.gov/state/index.php?sid=HI. Accessed 12 May 2023
3. U.S. Energy Information Administration, February 2023
4. Renewable Energy Procurement (Docket No. 2017-0352). https://puc.hawaii.gov/energy/renewable-energy-procurement/. Accessed 12 May 2023
5. Integrated Grid Planning for Hawaiian Electric (2018-0165). https://puc.hawaii.gov/energy/integrated-grid-planning-docket-for-hawaiian-electric-2018-0165/. Accessed 12 May 2023
6. Energy Equity and Justice (Docket No. 2022-0250). https://puc.hawaii.gov/energy/equity/. Accessed 12 May 2023
7. PUC FAQs. https://puc.hawaii.gov/faqs/. Accessed 12 May 2023
8. CPUC News and Outreach Office. https://www.cpuc.ca.gov/about-cpuc/divisions/news-and-public-information-office. Accessed 12 May 2023
9. FTI Consulting, Inc. Prepared for National Association of Regulatory Utility Commissioners: State Approaches to Intervenor Compensation (2021)
10. CPUC Low Income Oversight Board. https://liob.cpuc.ca.gov/. Accessed 12 May 2023
11. CPUC Disadvantaged Communities Advisory Group. https://www.cpuc.ca.gov/industries-and-topics/electrical-energy/infrastructure/disadvantaged-communities/disadvantaged-communities-advisory-group. Accessed 12 May 2023

Equitable Energy Ownership and Access Policy, Regulation, and Investment

Accelerating 100% Renewable Energy Plans

Steven B. Smiley(✉)

Energy Economist, 2019 Boren Ave #2910, Seattle, WA 98121, USA
smiley27@earthlink.net

Abstract. The focus of this paper is accelerating a ten-year 100% renewable energy plan for a city with a population of 15,000, Traverse City, Michigan. This plan demonstrates that a typical municipal electric utility can double and even triple its electric distribution with small changes to existing infrastructure– displacing natural gas, the largest greenhouse gas source in the city–beginning with a menu of solar photovoltaics (PV), thermal and electric storage, infrastructure upgrades and smart grid broadband applications. This is a conceptual plan, demonstrating how to cut greenhouse gases (GHG) 50% within three years. This plan addresses some of the impediments, and challenges conventional thinking about how to address the climate crisis in the developed world. The conventional thinking includes; that more grid transmission is required; that long permitting delays and independent system operator (ISO) queues will bottleneck progress; that electric battery storage should be a priority over thermal storage; that energy efficiency should be a priority over renewable energy; that renewable energy integration capacity in our utilities is limited; that natural gas should be a transition fuel; that electricity cannot out compete natural gas; that fuel switching to electrification is going to be expensive; that atomic energy is necessary; that there are not enough renewable resources or land; that the technology is not yet advanced and ready; that electric vehicle charging infrastructure will be constrained; that energy democracy and justice cannot be achieved without great expense; and that electric prices must rise to solve the climate crisis.

Keywords: Accelerating renewable energy · 100% renewable energy plans

1 Introduction

The focus of this paper is accelerating a ten-year 100% renewable energy plan for a city with a population of 15,000, with a preliminary goal of achieving 50% renewable energy penetration within three years, cutting greenhouse gas emissions by over 50%. This acceleration is possible with the application of the policy and project prescriptions outlined in the authors "Emerald City 100% Energy Plan" [1]. Emerald City is a stand-in name for Traverse City, Michigan in a northern humid continental climate, with 7800 heating degree days (HDD), latitude 44° north. The ten-year plan is built on the baseline energy profile of the city's electric and natural gas consumption. While this baseline data is from 2014, energy consumption has not significantly changed (See Fig. 1). Electricity is distributed by the city-owned municipal electric utility, with 2022 data indicating

D. Renné et al. (Eds.): ASES SOLAR 2023, SPE, pp. 155–167, 2023.
https://doi.org/10.1007/978-3-031-39147-7_18

12,969 metered electric customers, an electricity distribution of 329 million kW-hours per year, and an annual peak of just under 70 MWe. Natural gas, the primary heating fuel, is supplied by a state-wide independent private investor-owned utility (IOU). The energy distribution in the ten-year plan (see Fig. 2) targets the displacement of natural gas, vehicle fuels, and the phase out of the utility power pool ownership of coal-fired, and natural gas fired generation; utilizing a menu of solar photovoltaics (PV); wind; biogas; thermal and electric storage; infrastructure upgrades; and smart grid broadband applications.

This three-year plan to double electricity distribution with renewable energy is accomplished by instituting policies, infrastructure upgrades, smart grid broadband, solar electric PV, thermal and electric storage. With this foundation established, longer-term systems development can be integrated to further displace natural gas–including wind power, dedicated renewable energy sub-station capacity, and biogas combined heat and power (CHP). These systems are critical to meeting wintertime heating energy loads with electrification—the ten-year goal.

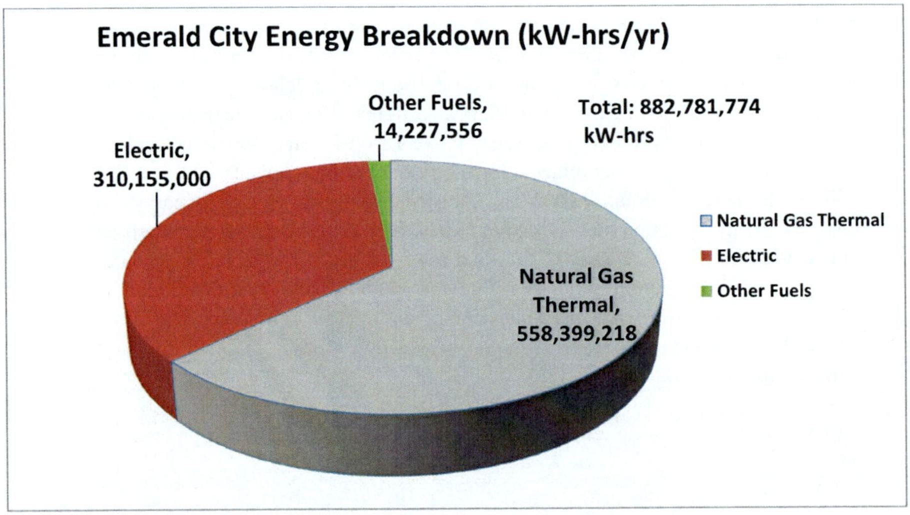

Fig. 1. Total energy breakdown, electric, natural gas, and other fuels.

Figure 3 illustrates the importance of the seasonal balance between solar and wind resources in this region, where low winter sun is offset by higher winter wind resources.

The goals of this three-year conceptual plan for accelerating renewable energy and abating greenhouse gases (GHG) include:

- Increasing the distribution of municipal electricity by a factor of two, from 329 to 600 million kW-hours per year—with added energy generated primarily from solar electric systems.
- Reducing consumer retail per kW-hour electric prices ten percent or more.
- Displacing the largest GHG source, natural gas, by over 50%.

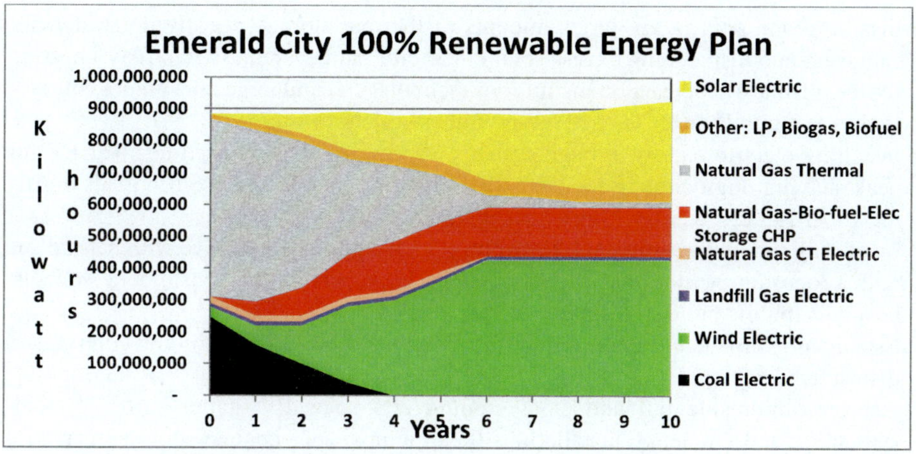

Fig. 2. Ten-year 100% renewable energy plan to displace natural gas and coal.

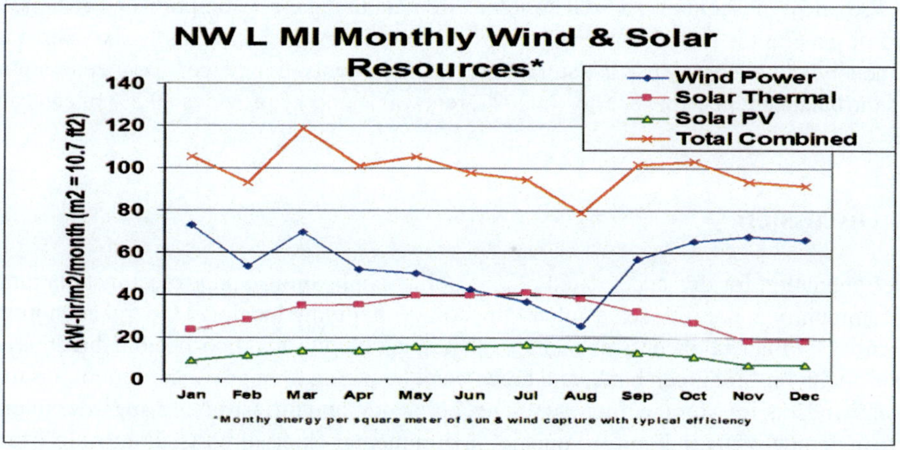

Fig. 3. Northwest lower Michigan wind and solar seasonal distributions

- Installing midscale community, cooperative and utility scale solar electric systems (3–100 MWe) of approximately 110 MWe inside the city boundaries and 90 MWe in the surrounding franchise areas.
- Eliminating interconnection delays, with no "Independent System Operator" (ISO) requirements and transmission upgrades.
- Instituting comprehensive electric rate structure reform with time-of-day pricing for all customer classes, providing lower cost, off-peak period electricity, eighty percent of the time.
- Selling electricity (mostly off-peak) at a price competitive with natural gas–and much lower than petroleum–for space, water heating and petroleum vehicle fuels—accelerating renewable energy competition.

- Installing and controlling large amounts of thermal storage, mostly hot water, for capturing and dispatching excess solar electricity (and eventually wind) as a heating source, primarily focused on institutions (schools, etc. and large consumers with, for example, ten million BTU–3 MWt–or higher boilers).
- Installing electric battery storage systems, especially for summertime solar PV for peak shaving, high cooling loads and EV charging, displacing electric loads during the 10 AM–5 PM weekdays peak period hours.
- Instituting municipal utility sales of hot water, at a price competitive with natural gas with a focus on institutional and commercial large heating load customers (schools, hospitals, public buildings, hotels, etc.).
- Establishing broadband smart grid infrastructure and controls to monitor, control, and dispatch electricity.
- Accelerating residential and small commercial solar PV, battery storage, fuel-switching, and efficiency installations (heat pumps, etc.) connected to a smart grid harmonization system, encouraged with marketing and incentives, including rebates, credits, full net metering, and on-bill financing.

Presently, most of the city's GHGs come from natural gas consumption and coal fired electric generation. Of the total energy use (not including vehicles), roughly sixty percent of the city's total energy use is natural gas, roughly twenty-five percent coal generation and the balance renewable energy. In 2022, renewable energy provides 29.5% of electric sales.

2 Discussion

The foundation for this accelerated plan is policy applications and project installations of community and utility scale solar with storage, applying local and federal economic incentives. Financial incentives include rebates, tax credits, and non-taxable direct payments–with on-bill, green bank, and green bond financing to improve the economics for solar, wind, energy storage, fuel-switching, efficiency, and infrastructure upgrades internalizing some of the heretofore unpaid environmental external costs. Following from the city utility energy profile (Fig. 1 and 2) it is apparent that an accelerated renewable energy plan must focus on commercial and utility-scale installations, with emphasis on community, cooperative and municipally owned solar and storage—all inside the municipal utility electric distribution system. Since natural gas is the largest source of greenhouse gas emissions in the community, fuel-switching to renewable electricity is a critical component of this plan. With rate reforms, off-peak electric energy pricing can out-compete natural gas and other fossil fuels for space, water, and process heating, and at a fraction of the cost of petrol for electric vehicle charging.

By placing the solar generation inside the utility distribution system there are no required applications or engagement with the mid-west regional electric grid independent system operator for new electric generation; in this location, the Mid-Continent Independent System Operator (MISO) which has an interconnection queue of two or more years. With the plan advocated here, no new transmission requirements are needed. Indirectly, this opens transmission capacity for others. This approach abates the need for transmission upgrades, demonstrating that calls nationally for massive transmission capacity is substantially overstated.

With mid-scale solar arrays, say 1,000 kWe and greater, the cost is roughly 4 cents per kW-hour. This is especially true when either the production tax/non-taxable direct pay credit (PTC) incentive of 2.6 cents per kW-hour or the 30% Investment (ITC) tax/non-tax direct pay credit financial advantages are included. We can safely assume the future solar energy value will be under 4 cents per kW-hour. Presently, the cost of natural gas for heating is between 4 and 5 cents per kW-hour, varying with the system efficiency (70%–90%) and the delivered price of natural gas (US$1.00 ± per CCF or therm). This cost comparison is for direct electric resistance heating, (domestic hot water tanks, baseboard electric, all other internal electric sources that heat such as lighting). With the application of heat pumps for space heating and domestic hot water (DHW), the value of electric energy for heating is more than doubled (cutting the heating cost of electricity 50% or more). For example, if off-peak electric energy can be delivered at 5 to 7 cents per kW-hour, the comparison value is 2.5–3 cents/kW-hour, contrasted with a natural gas price of between 4 and 5 cents per kW-hour. With these low prices, thermal and electric energy storage can be applied cost competitively to provide for the distribution of the solar energy both on-peak and off-peak. During on-peak periods (10 AM–5 PM weekdays) solar energy will directly off-set higher cost electricity, driving overall electric prices downward.

Mid-scale storage can be implemented at an accelerated pace. With a focus on mid-scale solar and energy storage, electric thermal storage (hot water) can be distributed primarily in conjunction with facilities that have large natural gas boilers, such as schools, hospitals, and public works, where 3 to 5 MW storage and heating systems (10–15 million BTU) can be installed. Variable-controlled electric boilers with associated hot water storage in the range of 3 to 5 MW can be pre-fed into existing natural gas boiler systems with little change in the mechanical systems. Such systems can also be utilized for electric power quality management, and balancing system loads. Ten or more of such 3–5 MW systems can provide 30 to 50 MW of storage capacity. In addition, thousands of existing electric DHW tanks can be utilized with smart grid controls–simply adding low-cost thermal storage with demand management. Starting with the larger loads, such controls can be tested and implemented under a new "smart grid" program with the city broadband fiber system, accelerating the plans. There are twenty-five public school facilities in the community, from elementary to high schools, and a community college that could be targeted for electrification, thermal and electric battery storage systems. There are dozens of such creative approaches and opportunities to accelerate this plan.

Proactively–to advance the implementation of a modern smart grid system–the municipal utility applied to the United States Department of Agriculture and received a $13.8 million low interest loan. The following Fig. 4 illustrates the nature of the smart grid system.

Mid-scale electric battery storage should be distributed by substation circuit transformers sized to address, primarily, summer peak period loads, installed in combination with distributed community solar arrays. With the planned excess energy put in storage, thermal and electric energy can be delivered at off-peak electric rates, approximately 5–7 cents per kW-hour for a "high load factor" customer. These off-peak periods include

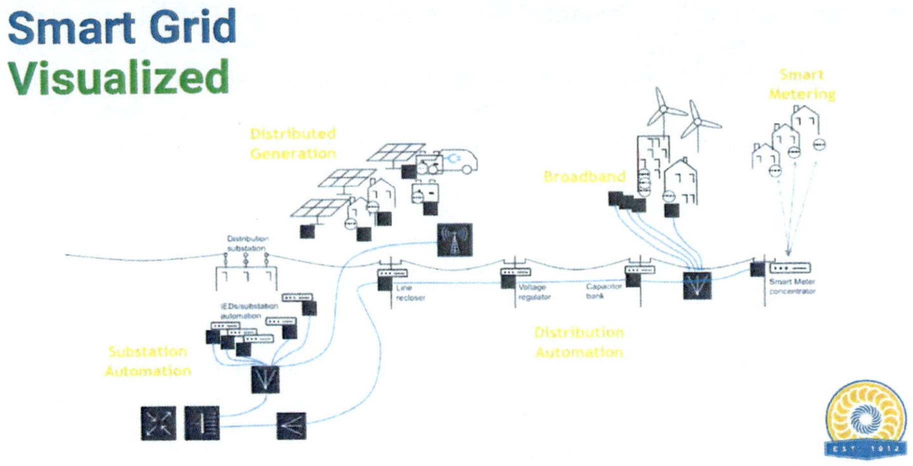

Fig. 4. Smart Grid Visualized

all weekends (65 h), holidays and all but seven hours each weekday. Eighty percent of the hours in a typical week are off-peak.

Presently, high load factor rates are only offered to large primary industrial, institutional, and commercial customers with the "Primary Service-High Load Factor (Rate I1)", which has a high monthly fixed fee ($200/mo.), a peak period capacity (demand) charge of $14.05, an off-peak energy charge of 5.06 cents per kW-hour, and on-peak charge of 6.26 cents per kW-hour. Customers with load factors over 90% receive a 5% credit on the total amount billed; load factors over 70% and 80% receive a 3% and 4% credit respectively. The high monthly fixed fee is typical for primary rate industrial customers with large transformers and should not apply to other rate classes. If the capacity charge of $14.05 per kW of on-peak demand can be offset with "smart controls", demand management and solar PV, average retail electric prices can approximate 5–7 cents per kW-hour, for these high load factor customers.

A load factor is calculated by dividing the energy of a time period, over the peak multiplied times the time period. So (for example) the annual energy of 329,000 MW-hours divided by the annual peak of 70 MW multiplied times 8,760 h per year determines the load factor–or mathematically, $(329{,}000/(70 \times 8760)) \times 100 = 53.6\%$.

With an existing summer peak capacity of roughly 70 MW's, the present overall city load factor is 54% which means that theoretically 46% of capacity remains available to distribute more energy. In principle, there exists adequate electric distribution capacity without significant infrastructure upgrades. At a 100% short-term load factor, demonstrated when there is a peak load of 70 MW during a hot summer day, all consumers are served, suggesting that the existing distribution system can perform without significant infrastructure upgrades. Of course, there must be some normal maintenance and infrastructure upgrades–transformers, distribution sub-stations, smart controls, SCADA systems, power quality improvements and circuit breakers.

Prudently, this well managed and serviced utility completed a "2021 System Load Study and Analysis" [2] which describes the nature of the system, electric loads, recommendations, and budgetary costs for improvements. System upgrades are proposed with a total cost estimate of $19 million, including capacity increases at two substations. There are federal subsidies and low interest guaranteed loans available for these upgrades. The local grid consists of five distribution substations, ten substation transformers rated from 17.9 MVA to 37.3 MVA and twenty-eight 13.8 kV distribution circuits. Table 1 (below) shows the nature of this distribution system, and the electric loads demonstrate there are ample opportunities for integrating mid-scale solar PV, thermal and electric battery storage throughout the community.

The box on the left in Table 1 is data provided from the study, with transformer ratings and summer peak loading, with a percentage of full load capacity, columns (b) and (c). As seen in column (b) below, eight of the ten substation transformers have peak loading well under 45%, which is designed to provide a safety factor should a substation transformer circuit fail, enabling back up from adjacent transformers and circuits. The highest peak loaded transformer is Parsons #2 at 51.7%.

The box on the right in Table 1 is an extrapolation by the author of the data from the system load study. Column (d) results from subtracting the summer peak, for example, Barlow #1 of 6.95 MVA (b), from the rated 22.4 MVA transformer (a), resulting in a net of 15.45 MVA. Column (d) shows the present excess capacity over the summer peak on each transformer in comparison to the 100% rating on each transformer, in contrast to the 45% loading. The sum of this, net of 100% capacity, totals 163.75 MVA.

Column (e) is the average load compared to the peak load, assuming 40 MW (MVA) is the annual average, and 70 MW (MVA) the annual peak. While the actual annual average is roughly 35 MW, 40 MW is assumed for a margin of safety. For example, the Barlow #1 transformer summer peak of 6.95 MVA multiplied times (40/70) results in an average loading of 3.97 MVA shown in column (e)–an average percentage load of 17.7% shown in column (f). This is well under the 45% limit with excess capacity available of 27.3%. Column (h) shows the Barlow #1transformer maximum target of 45% capacity of 10.08 MVA, which is 6.11 MVA over the average of 3.97 MVA.

This data indicates that the combined distribution transformers are loaded, on average, at 19.5%, (column (f)) –with a low of 9.3% and a high of 29.5%. Figure 5 presents a graphic representation of the ten transformer percentage loads. With the total transformer capacity of 244.8 MVA, the sum of column (a), a 45% loading equals 110 MVA, shown as the sum of column (h). A simplified examination of this data suggests, on average, 110 MVA of combined loads are possible without any planned or new capacity expansions. This transformer loading is assumed to be uncontrolled without demand management, energy storage or smart grid applications.

This average 19.5% loading factor suggests there is double and possibly triple the existing distribution of electrical energy—an amount, which is speculated to exceed 900 million kW-hours a year. This is what is required to substantially eliminate natural gas in the local energy system.

Detailed technical and engineering analysis is required to determine what the full potential is for interconnecting solar PV with storage on the distribution and customer side of the substations. Studies are needed to determine what are the load impacts if solar

PV is installed on the distribution and customer side equivalent to the average or peak loading. For example, column (b) indicates a summer peak load of 6.95 MVA on Barlow transformer #1. If 6.95 MVA of solar is installed on the customer side of the substation (community or utility owned solar), what is the impact on Barlow transformer #1? Does the transformer load become net zero? If the transformer load does become net zero, what are the implications for adding more solar or other renewable generation sources on the distribution substations? How will this be managed by a smart grid system? How is this situation different from a typical residential solar PV grid interconnection? What is the optimal thermal and electrical battery storage sized to meet summer peak loads and winter heating loads on the distribution substations? How should energy storage investments be shared between the customer and the utility, as part of HVAC electrical and mechanical upgrades or part of the entire utility system? Is there enough distribution capacity, as speculated by the author, to heat the entire city with the proposed electric solar, wind and CHP capacity? What electric distribution controls and safety breakers are required, and at what cost? Many questions remain to be addressed by utility distribution engineers.

Table 1. Transformer loading – peak summer and average

LOAD STUDY DATA FROM ENGINEERING REPORT					SMILEY EXTRAPOLATION OF DATA					
		(a)	(b)	(c)	(d)	(e)	(f)	(g)	(h)	(i)
Substation	Trans	Rating (MVA)	Summer Peak Loading (MVA)	% Max	Net of 100% (MVA)	Avg Loading at 40/70 (MVA)*	% Avg	Net of 45% Safety Factor**	45% Avg Load (MVA)	Avg Net Load at 45% (MVA)
Barlow	#1	13.4/17.9/22.4	6.95	31.0%	15.45	3.97	17.7%	27.3%	10.08	6.11
Barlow	#2	13.4/17.9/22.4	7.54	33.7%	14.86	4.31	19.2%	25.8%	10.08	5.77
Cass	#1	13.4/17.9	7.57	42.3%	10.33	4.33	24.2%	20.8%	8.055	3.73
Cass	#2	13.4/17.9/22.4	11.41	50.9%	10.99	6.52	29.1%	15.9%	10.08	3.56
Parsons	#1	13.4/17.9	7.00	39.1%	10.90	4.00	22.3%	22.7%	8.055	4.06
Parsons	#2	13.4/17.9/22.4	11.58	51.7%	10.82	6.62	29.5%	15.5%	10.08	3.46
Hall	#1	22.4/29.9/37.3	9.57	25.7%	27.73	5.47	14.7%	30.3%	16.785	11.32
Hall	#2	22.4/29.9/37.3	11.03	29.6%	26.27	6.30	16.9%	28.1%	16.785	10.48
South	#1	13.4/17.9/22.4	3.63	16.2%	18.77	2.07	9.3%	35.7%	10.08	8.01
South	#2	13.4/17.9/22.4	4.77	21.3%	17.63	2.73	12.2%	32.8%	10.08	7.35
		244.8	81.05		163.75	46.31	19.5%	25.5%	110.16	63.85

*40/70 MVA is annual averag 40 MVA compared to summer peak 70 MVA multiplied times the summer peak MVA load.

**45 percent is the recommended transformer loading factor to provide 100% backup connecting to adjacent transformers with circuit failures

965,001,600 kW-hours/year with average loading of 110 MVA at 45% substation safety factor! This is 3 times the present annual consumpion!

Distribution substation infrastructure, with associated solar and storage additions, can be upgraded as part of recommended improvements. Under the federal Inflation Reduction Act (IRA) there are 30% investment direct pay credits now available for infrastructure, assuming the renewable energy system is sized less than 5 MW per installation. This applies to interconnection property associated with the installation of renewable energy property with a maximum net output of not greater than 5 MW-AC to provide for the transmission of distribution of the electricity produced or stored by such property,

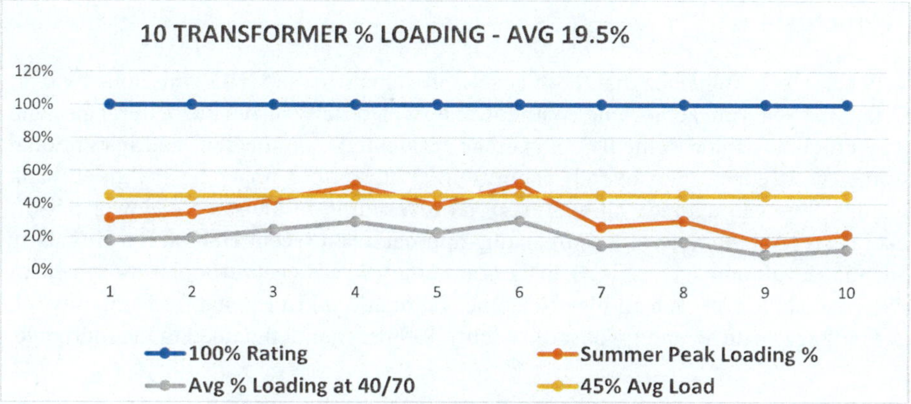

Fig. 5. Ten transformer percentage loads

and which are properly chargeable to the capital account of the taxpayer. The direct pay option allows non-taxable entities to directly monetize certain tax credits. The provisions apply to nonprofits, a state or political subdivision thereof, and such applicable entities can elect to be treated as having made a tax payment equal to the value of the tax credit they would otherwise be eligible to claim. The entity can then claim a refund for the excess in taxes they are deemed to have paid. The option effectively makes this tax credit refundable for these non-profit entities.

With ten transformer circuits that includes community solar PV arrays under 5 MWe per transformer, with an associated thermal and electrical storage system under 5 MWt, this approach can provide nearly 50 MW of solar and storage–roughly 73 million kW-hours per year of solar generation–under community, cooperative and municipal ownership with 30% direct payments for substation infrastructure improvements.

One possible solar/storage application is a 3 MW (10 million BTU) electric resistance boiler, with variable controlled operation, installed in conjunction with large thermal storage, and a multi-megawatt solar array, that can be dispatched via the smart grid in accordance with the transformer loads—heating a school (for example), and displacing large amounts of natural gas.

The balance of the required solar electric systems can be installed according to the location of sites on each transformer circuit that can accommodate solar PV in ranges from 10 MW to 20 MW totaling roughly 110 MW. This would require roughly 550 acres (224 ha) inside or adjacent to the city, approximately ten percent of the city land area of 8.7 sq. Miles (22.5 km^2), where over 30% of the land is sacrificed for automobiles. In the "Emerald City 100% Renewable Energy Plan" eleven solar locations throughout the city are listed that might provide the necessary space, including the airport, schools, college, large parking areas and rooftops. The additional required solar, targeting a total 300 million kW-hours per year can be met with a utility scale system of approximately 90 MW on the required 450 acres (180 ha), installed in surrounding rural townships where the utility has franchise agreements. Eventually, new higher voltage collection substations in the distribution system will be required for planned utility scale solar and wind plants (roughly 16 wind turbines) during the ten-year development period.

3 Conclusions

This concept plan indicates that these accelerated goals to cut GHG emissions by over 50% in three years are achievable. With lower cost electricity displacing natural gas, and energy efficiency improvements, an average residential, commercial, and institutional customer will reduce their overall energy bill. With on-bill financing (or green bank financing), one can upgrade all energy systems resulting in the same or lower out-of-pocket costs. An example of this financing approach is a recent Habitat for Humanity residential development (see Fig. 6) in the community where a net zero solar PV system of 8 kW was installed on each all-electric home and financed in a Habitat for Humanity 30-year mortgage, with an energy cost of roughly $35 per month included in the mortgage.

Fig. 6. Habitat for Humanity "Depot" Community, with 8 kW solar all electric homes.

Residential full net metering should be maintained and increased without capacity limits as argued in my 100% plan. Such a program will have little impact on overall electric rates (one percent plus or minus by my calculation) due, in part, to limited opportunities for residential installations. With limited roof space and tree shading (solar cooling), at most, twenty percent of homes may have reasonable solar access. This points to the importance of community, cooperative and public utility solar systems so all citizens have an opportunity to invest in solar.

Energy efficiency marketing initiatives must be continued and expanded with on-bill financing, rebates, and tax credits—including incentives for energy storage, EV charging and controls. Fuel-switching incentives should be provided to change from natural gas to electricity. While most residences and businesses may not have space for a solar PV system, all can and should electrify to displace natural gas.

Transparent, publicly available information provided by public power municipal utilities is critical to addressing the climate crisis. Open monthly (or more frequent) board meetings with management and staff, directed by local board members, with all aspects of utility operation and planning available to the public, provides the knowledge and understanding to help citizens collaborate and participate in addressing the climate crisis in a democratic manner.

Ownership of municipal and citizen renewable energy plants and infrastructure is critical. Local ownership matters. With the recently adopted "direct pay" option provided by the federal Inflation Reduction Act for non-profits, there are few financial excuses to depend on power purchase agreements (PPAs) or carbon offsets for the application of renewable energy. PPAs and carbon offset projects significantly reduce the ability of cities and citizens to control their energy systems. On-bill financing with community or cooperative solar projects, with local ownership also financed by citizens, and public utility green bonds, provides opportunities for low-income and fixed income households to participate. Having an electric meter and bill should qualify any citizen, including low income and fixed income, to participate. These related energy justice issues have been addressed in the author's paper "An Energy Democracy Evaluation for 100% Renewable Energy Policies and Projects" [3].

With increasingly high load factors and displacement of natural gas over the ten-year period, as the utility doubles and eventually triples electric distribution, we will see lower marginal costs associated with these increased electric sales, providing for per kW-hour price reductions, easily ten percent or more.

When a locally owned electric utility can compete head-to-head with renewable energy against a major outside energy provider that is "selling" greenhouse gas emissions, accelerated change is possible. When one unit of natural gas is displaced, the money is kept in the local community with an economic multiplier effect, at minimum, a factor of two, traditionally shown by economists. With annual natural gas expenses of $20 million, there are significant economic benefits available to the community.

With the additional renewable energy generated and consumed inside the distribution system, this avoids any transmission and independent system operator requirements. Fossil fuel power purchase obligations delivered from outside the distribution system should be phased out over time, as quickly as possible, abating loads on the regional transmission system, freeing up transmission capacity for others, at no cost.

To accomplish this goal, roughly 200 MWs of solar electric generation needs to be planned and installed at the community and utility scale, with associated thermal and electric energy storage. Residential and small commercial solar installations, including storage and fuel-switching from gas to electric, should be encouraged with utility assistance, full net metering, rebates, and on-bill financing.

With the primary target of displacing natural gas heating, hot water storage and heating system retrofits are critical. Since most large commercial and institutional heating systems are based on hot water boilers it is a simple mechanical retrofit to pre-feed the existing systems with low-cost electric boilers along with hot water storage. Thousands of existing and new electric hot water heaters can be dispatched to provide hot water storage, with or without heat pump units. Heat pump direct exchange units can be retrofitted into heating and cooling force air systems without a major retrofit expense, especially when financial incentives such as rebates, tax credits and on-bill financing are provided. Natural gas heating units that are aging out or in need of expensive repair can be quickly recycled out of the system. When commercial wind energy, with a longer lead time, is eventually provided, the necessary infrastructure and thermal storage can be in place to deliver the low-cost wind energy, especially in the winter, balancing seasonal solar generation. On a cold windy night wind energy will have a marginal cost under two cents a kW-hour, providing enough energy to substantially heat the city.

With a peak period from 10 AM to 5 PM weekdays, solar PV with electric battery storage will play a crucial role for consumption and demand management, especially during the summer months. Smart grid management will be critical to control and optimize solar generation. Distributing mid-scale solar arrays throughout the city can abate the effects of clouds that can rapidly vary solar generation from one array to another. Smart grid controls can help manage the changing electric generation and loads. Prudently, utility has been undertaking the implementation of a city-wide broadband system for years that provides the backbone for the implementation of a smart grid system.

With comprehensive rate reform, establishing off-peak pricing for all customer classes with demand management and smart grid harmonization, low off-peak electric energy prices can out-compete natural gas—providing a major competitive force for change.

The city utility may consider selling hot water to large commercial customers and institutions such as schools, providing a utility financed energy service. An analysis should be conducted to judge the feasibility of this service.

It is assumed that with extremely low off-peak electric fuel prices, rebates and incentives, EV uptake will naturally accelerate. Electric vehicles and charging infrastructure will continue to be phased in naturally as demand increases. With low off-peak electric prices and incentives, with declining EV prices, increasing vehicle models available and fuel priced at under $1 per petrol gallon, EV's will become ubiquitous. Given that off-peak hours occur 80% of the time, including 65 off-peak weekend hours, there is adequate capacity on the distribution grid for scheduling and controlling vehicle-charging off-peak—especially utilizing EV chargers that don't require higher amperage upgrades.

With 2022 utility services operating revenues of $31.5 million per year–with distribution of roughly 329 million kW-hours per year–the average rate is under ten cents per kW-hour. Approximately 60% percent of expenses include power purchases from outside generation, averaging roughly seven cents/kW-hour. In time, local ownership of renewable energy generation can displace outside energy purchases for electricity and especially natural gas, and with solar and wind power costs roughly 4 cents per kW-hour, retail rate reductions will be possible.

The information and load data presented by the system study suggests that electric distribution from solar PV can account for an increase from 329 million to 600 million kW-hours in three years. "The 2021 System Load Study and Analysis" which provides specific electric loading on the distribution substations and transformers indicates that average loading varies from ten percent to thirty percent. At peak loads, the range is from 16% to 50%, depending on the specific substation transformer. Under such conditions, even without the planned capacity increases on two substations it is speculated that electricity distribution can triple while maintaining the 45% loading safety factor. The challenge to distribution electrical engineers is designing and installing the appropriate load management, power quality equipment and controls, and bi-directional safety breaker systems.

This concept plan is most applicable to municipal public power utilities, especially where the natural gas supplier is not locally owned. This situation provides direct competition and the additional economic advantages of displacing natural gas delivered from outside suppliers. While municipal power utilities represent a small portion of the national grid, such plans can demonstrate the potential to others. All municipal utilities are bordered by IOU's or electric cooperatives, and–when their consumers see the potential for cost savings, renewable energy and GHG reductions, added pressure will be put on them to advance their plans to address the climate crisis in a similar manner.

Public power municipal utilities operate under local control and management with significant flexibility to institute policies and projects without delay. New electric rates and policies can be instituted in a matter of months, in contrast to the long lead times for IOUs under state regulation. It is not unrealistic to plan, develop, permit, and install solar electric systems within a three-year time frame. This demonstrates it is feasible to accelerate the implementation of renewable energy from added solar electric systems with energy storage and smart grid harmonization, increasing distribution from 329 to roughly 600 million kW-hours per year, cutting GHGs over fifty percent.

References

1. Smiley, S.B.: Emerald City 100% renewable energy plan. In: Proceedings of the American Solar Energy Society National Conference (2020)
2. Traverse City Light and Power, Traverse City, Michigan: 2021 System Load Study and Analysis. GRP Engineering, Inc., Petoskey, Michigan
3. Smiley, S. B.: An energy democracy evaluation for 100% renewable energy policies and projects. In: Ghosh, A.K., Rixham, C. (eds.) Proceedings of the American Solar Energy Society National Conference: ASES SOLAR 2022, pp. 296–307. Springer, Cham (2022). https://doi.org/10.1007/978-3-031-08786-8_32

Exploring New Deployment Tools for Accelerating Equitable Solar Access to Low-and Moderate-Income Households

Elvis Moleka(✉)

Groundswell, Washington DC 20003, USA
elvis.moleka@groundswell.org, emoleka@lekaresearchinstitute.com

Abstract. The Accelerating Low-Income Financing and Transactions (LIFT) for Solar Access Everywhere project researched and gathered data on 453 community solar projects in the United States, that dedicated some portion of their system capacity to Low-and-Moderate-Income (LMI) households. This paper discusses new community solar deployment tools, strategies for implementing financial best practices, and other key insights utilities and solar developers can leverage to improve inclusive solar access to LMI communities. The research reinforces the need for software development tools and programs targeted at LMI households – especially those severely impacted by high energy burdens. This developmental need is clear because although weatherization policies are necessary, they are not sufficient to address systemic root causes of energy burden, as there are more eligible households than available program space within federally mandated energy assistance programs. Incentives on energy efficiency programs – including cost-effective financing mechanisms, at the local and state levels – are needed to make improvements that effectively address energy burdens for renters in multifamily housing and single-family residential homes.

Keywords: Community Solar · Energy Efficiency · Equitable Solar Access

1 Introduction

Equitable solar access has long been an issue of concern when studying the prevalence of high energy burdens, systemic poverty including the impacts of redlining, and environmental injustices in LMI communities (Brown et al. 2020; Moleka 2021). However, alarmingly, despite energy abundance in the United States, coupled with the propagation of bill-payment assistance programs, and weatherization policies, low-income households continue to pay high energy bills while their environmental, social, and economic conditions erode (Moleka 2021). Additionally, many low-income households live in older homes, marked by structural issues, such as poor insulation, inefficient HVAC systems, leaky roofs, inefficient and sometimes oversized appliances which increase energy costs (Reames et al. 2021; Liddell and Morris 2010; Hernández and Bird 2010). For most LMI households, energy efficiency is more of a "dream" than a "reality." Against this background, this paper sought to assess opportunities for expanding energy programs

D. Renné et al. (Eds.): ASES SOLAR 2023, SPE, pp. 168–177, 2023.
https://doi.org/10.1007/978-3-031-39147-7_19

that offer the greatest hope to reduce energy burdens, to increase equitable access to clean energy through deployment of new energy technologies, and to expand development of community solar projects serving LMI households in the U.S. Energy burden is defined as the percentage of household income spent on electricity and heating services. That is, the total household utility expenditures for heating and cooling as a percentage of total income (Chai et al. 2021).

Table 1. Distribution of Project Capacity and Regulatory Framework by State.

State	Community Solar Enabling Legislation	Avg. LMI Project Capacity KW-AC	Avg. Potential # LMI Subscribers	Avg. LMI Customer Savings %	Potential # LMI Subscribers
CA	Yes	2,643	3,175	43	9,526
CO	Yes	157	65	50	4,167
CT	Yes	103	41	10	123
DC	Yes	350	132	98	6,220
FL	No	1,742	697	13	10,451
IL	Yes	1,296	468	54	6,080
MD	Yes	633	251	27	8,021
MA	Yes	560	194	36	12,982
MN	Yes	527	179	6	1,254
NH	Yes	57	23	81	117
NJ	Yes	661	264	20	38,338
NY	Yes	1,090	378	91	3,404
NC	Yes	25	10	100	10
OR	Yes	206	82	25	2,552
RI	Yes	472	100	12	100
TX	No	364	92	37	643
VT	Yes	61	45	54	90
WA	Yes	75	27	68	27
Overall Averages		**562**	**230**	**39**	**104,105**

The LIFT for Solar Access Everywhere research supports a large and growing body of evidence concerning the unjust realities of the clean energy transition in the U.S. (Bohr and McCreery 2020; Li et al. 2014; Buylova 2020). This inequitable transition exists in part because relative to demand, not many low-income households are able to benefit from energy-efficiency programs that are designed to reduce economic hardship and poverty (Hernández and Bird 2010). Therefore, the LIFT research hypothesis was conceived to deliver precise and thorough insights into the specific mechanisms that enable equitable access and customer experiences with solar energy programs that serve them. According to Kontokosta et al. (2019), while energy efficiency programs are known to successfully address energy burdens, "a growing body of research shows that [Black, Indigenous, and People of Color] BIPOC and low-income communities experience disparate access to residential energy-saving programs and other energy efficiency upgrades." Answering this call to action, LIFT collected and analyzed data on existing community solar and energy efficiency projects, as well as those in development, across seventeen states, and the District of Columbia, as shown in Table 1.

Most states analyzed have legislations that enable the development of community solar projects. Except in Florida and Texas, where no mandates exist for community solar development, but projects exist either through pilot programs or directly through utilities (Greco and Archbell 2022b). As a major contribution to the literature, the LIFT Toolkit for community solar development is the first in the industry to thoroughly design a solar energy technology to encourage developers in their design of clean energy programs that serve LMI communities.

2 Literature Review

Energy is a necessity that affects every aspect of society including housing, education, and mobility (Brown et al. 2020). However, energy poverty is unevenly distributed in the U.S., and LMI households spend twice as much of their income on electricity bills as compared to their non-LMI counterparts (Heindl 2015; Mohr 2018; Kontokosta et al. 2019; Ray et al. 2019; Moleka 2021). Similar findings on energy poverty and rising fuel energy prices have been noted in the United Kingdom (Liddell et al. 2012) and Australia (Chai et al. 2021). Studying the prevalence of energy poverty in the U.S. between 1999 and 2017, Bohr and McCreery (2020) showed that LMI households were more likely to transition to economic hardship and poverty as compared to non-burdened households. In the same vein, Hilbert and Werner (2016), studied energy poverty for New York households, and concluded that energy poverty results from low-income households living in homes deprived of basic energy needs, such as a lack of electric heating or modern cooking appliances.

According to Elnakat et al. (2016) and Buylova (2020), most low-income community census tracts are subject to high energy burdens, high climate pollution, and environmental hazards. Therefore, a plethora of studies have examined the health impacts of high energy burdens on LMI households, such as aggravated respiratory and mental health issues (Reames et al. 2021; Liddell and Morris 2010; Hernández and Bird 2010). Additionally, studies have linked high energy burdens to structural issues and housing inefficiencies, caused by leaky roofs, poor insulation, and inefficient HVAC systems (Bohr and McCreery 2020; Brown et al. 2020). Yet further, other studies have attributed high energy burdens to income disparities and demographic imbalances, with African American households being more vulnerable than White, Asian, and other minority group households (Wang et al. 2021; Chai et al. 2021; Brown et al. 2020). Meanwhile, utility rates amount to a lesser proportion of the variation in energy burdens among LMI residential households (Moleka 2021; Chai et al. 2021; Bohr and McCreery 2020; Mohr 2018; Wright 2004). The Ishikawa/Fishbone diagram (Fig. 1) is a summary of the root causes of high energy burdens for LMI households.

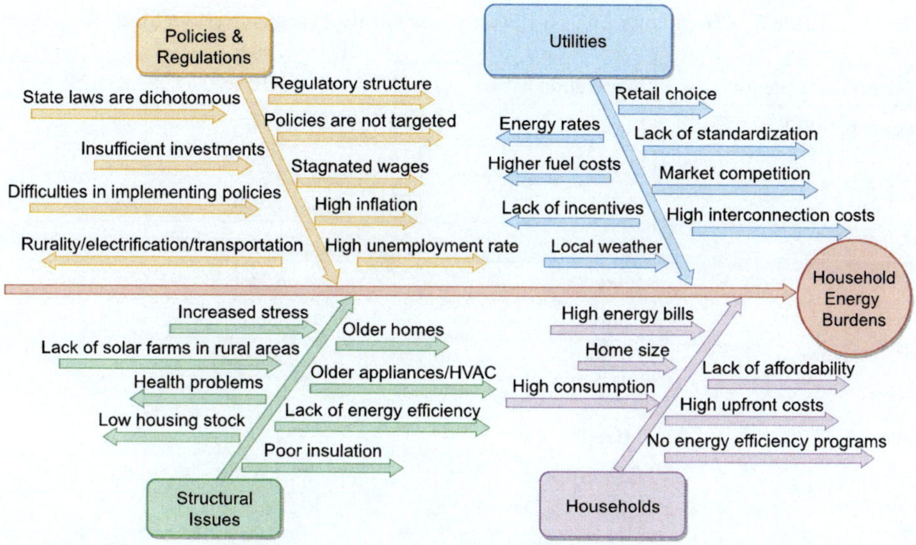

Fig. 1. An explanation of the root causes of energy burdens for LMI households.

As shown in Fig. 1, high energy burdens are the result of structural issues, poor implementation of regulatory policies to scale community solar projects, as well as utility policies and households' actions. Energy efficiency policies – especially those related to energy equity for targeted vulnerable LMI populations – are necessary to create an inclusive and equitable clean energy economy for all.

3 Data and Software Development Methodology

LIFT developed LMI characteristics for data collection by identifying 453 projects across the U.S. that dedicated some portion of the system capacity to LMI households – mostly represented by Investor-Owned Utilities (IOUs), while Municipally-Owned Utilities (MOUs) and Electric Cooperatives (ECs) together, represented just over 6% of the total system capacity in our sample (see Table 2). This represents the sad reality of the U.S. solar industry that policy regulations, do not target improvements to rooftop or ground solar installations by MOUs and ECs. Rather, the solar industry is dominated by large IOUs, crowding out the efforts of ECs and MOUs in the just energy transition movement. Therefore, the LIFT Toolkit was conceived as a framework for spurring development of community solar projects across regulatory environments, and all three utility types, centering development in tribal and BIPOC communities.

3.1 Design and Construction of Logic Models

As the LIFT dataset is one-of-a-kind, so too is the Toolkit itself; as no similar resource exists for developers, utilities, and financiers, to promote favorable community solar program design decisions, that reduce energy burdens. What makes the LIFT research

Table 2. Community Solar Installations by Utility Type and System Size

System Size Category	Installation Type	ECs	IOUs	MOUs
Less than 500 KW	Mixed	29	514	
	Canopy	170	1,691	
	Ground	951	1,529	64
	Roof	21	15,616	802
500 KW to 1,999 KW	Mixed		763	
	Canopy		1,602	
	Ground	2,935	28,348	
	Roof		38,194	
More than 2 MW	Mixed		1,204	
	Canopy		1,071	500
	Roof		29,968	
	Ground		119,586	8,224

different from other research on emerging tools and software developments is the fact that the Toolkit is based on an astute logic model for both the project search map, the "Project Map" – integrated with MapBox and geocode functionalities for each of the 453 projects – and the best practices builder, the "Project Optimizer" (see Fig. 2 and Fig. 3).

When a user clicks on a project pin on the Project Map or enters elements of their project in the Project Optimizer, algorithms will determine a response and output for that user. Optimization allows the user to change input fields (drivers) to achieve the desired level of financial performance, including best practices for their project. The Toolkit also includes a library of LIFT-generated and LIFT-curated resources for users to learn more about how to best serve LMI communities through community solar – as well as resources for outreach and community engagement. Like any software development product, user acceptability testing (UAT) was performed to ensure that the Toolkit satisfies the non-functional and functional requirements. The UAT tests (Alpha and Beta testing) were conducted for the primary user audiences of the Toolkit, such as solar developers, investors, utilities, municipalities, and researchers, who may leverage the emerging energy efficiency technology to reduce high energy burdens for LMI households.

With the Project Optimizer, user outputs are generated directly from fields the user enters for their project data – delivering projects with similar data – or derived from key findings within innovative projects – case studies and their financing best practices – as well as external resources – pulled from a curated library. Algorithms were defined for each type of output to result with the most applicable combination for the user's proposed project. Similar projects delivered are those with the most "hits" on the driver value matches between the proposed Project Profile – input criteria – and existing projects in the database. Each Best Practice has a set of driver values that must be matched, in

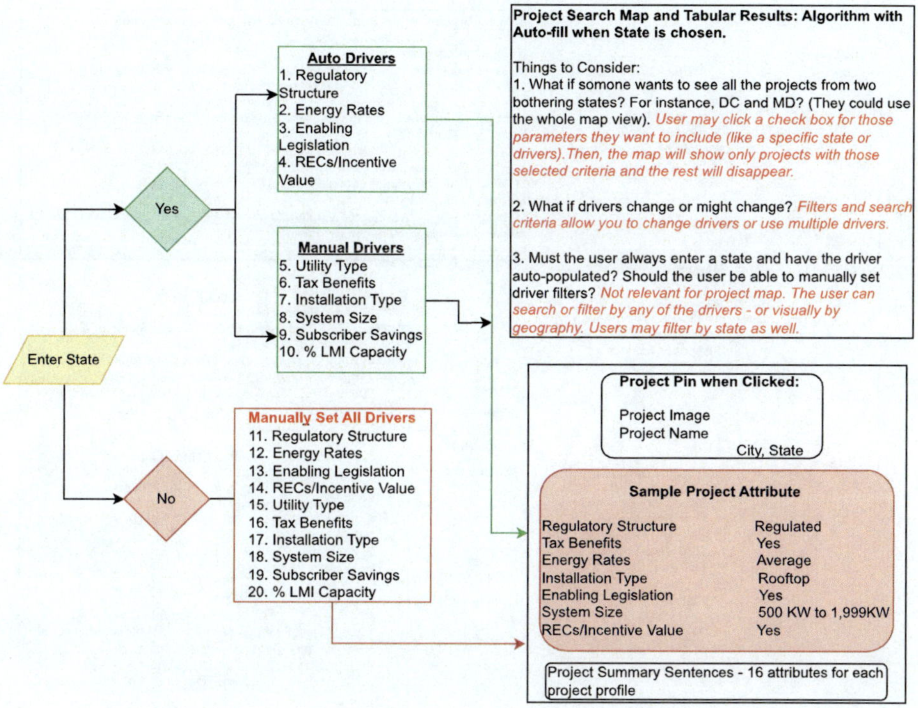

Fig. 2. Methodology Supporting a Filtered Search of Community Solar Projects.

a 1:1 value match or a combination. For example, if the filter for Tax Treatment = No, THEN the Best Practice "Tax Equity Investment" is delivered to the user, alongside case studies. Each case study has a set of best practices mapped to the drivers. Resources are also mapped to the set of driver values as best practices are.

3.2 Results of the LIFT Toolkit Functions

The LIFT Toolkit is designed for use by utilities, energy service providers, and financiers or investors as an educational and decision-making tool, to rapidly scale project models that optimize LMI inclusion – and maximize real household savings, thereby reducing energy burdens. The LIFT Toolkit is structured in three key functions to optimize equitable solar project deployment:

1. The Solar Project Map/"Explore" page: A map of the U.S. showing state border lines and 'pins' designating each of the 453 community solar projects that serve LMI residential customers, and links to the data collection methodology.
2. The Project Builder/"Design" page: An interactive page displaying ten 'drivers' – solar project characteristics – as drop-down options, based on Fig. 3. The page features directions for the User to explore 'building' a community solar project that serves LMI households, including finance-related best practices.

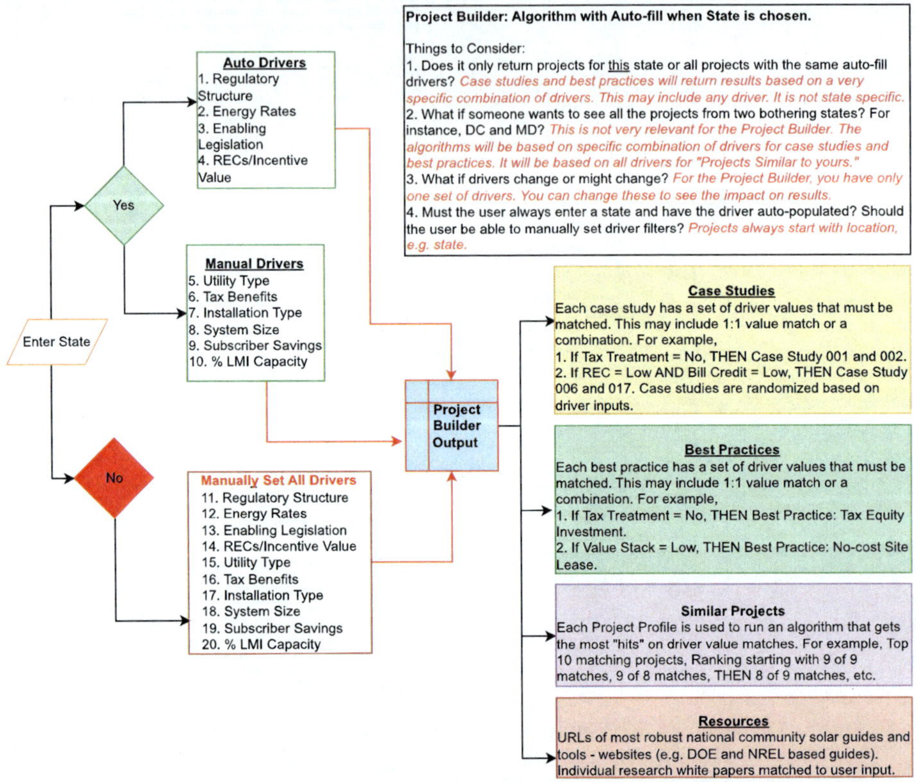

Fig. 3. Project Optimizer (Decision Support Tool) Logic Builder: Methodology to Match Best Practices to Project Builder Outputs.

3. The "Learn" Page: A page providing embedded LIFT, and external resources to support Users' development and access to LMI community solar resources. The page is divided into three sections:
   - LIFT Resources: With all benchmarking and final research reports (ten total), all twenty-three case studies and the list of project financing Best Practices.
   - External Resources: A tailored list of thirty-eight linked External Resources, organized into six topics – Advancing Equity, Basics of LMI Community Solar, Community and Consumer Resources, Design and Financing Models, National Level Datasets, including Policy and Market Transformation.
   - Glossary: Key LMI community solar terminology.

4 Results of Customer Experience and Solar Project Finance

Leveraging the insights from concurrent research on equitable access to community solar, the LIFT Solar Everywhere Research project understood primary program design drivers of equitable community solar programs – policy mandates, financial incentives, equity requirements, and voluntary utility-led programs – and key subscriber outreach features – subscriber eligibility, customer outreach, partnership outreach, messaging, and communication (Greco and Archbell 2022a). Recommendations in other reports on customer outreach, messaging and communication do propose an increased emphasis on the importance of a customer's experience, but still center the developer or administrator's perspective (Heeter et al. 2021). To address the limited polling of real community solar subscribers' experiences, LIFT used a human-centered design research approach, and surveyed 17 of the 453 community solar projects in the U.S., to gain insight into how LMI subscribers feel about community solar and the programs that serve them (Greco and Archbell 2022a). Subscribers indicated that the most motivating issue to participate in their program was not savings but "helping the environment."

Interestingly, these findings represent a noticeable shift in customer motivation from previous studies. In 2015, Smart Electric Power Alliance (SEPA) and the Shelton Group found "saving money" to be slightly more important for subscribers than "helping the environment." From the LIFT research, saving money is still important to all community solar subscribers, but the relationship between savings and satisfaction level is not as clear as previous research may have indicated (Greco and Archbell 2022a). LIFT also found that "35% of subscribers stated that savings was not clearly communicated to them." It therefore implies that customers who were promised high savings levels may not have received transparent and readily available information on what "savings" represent, and how they can be calculated.

Acknowledging that programs must be intentional in their design and chosen financing models to ensure savings flow down to LMI customers, LIFT examined community solar project financing separately from customer experience. Other research initiatives have studied the correlation between community solar project financing and equity, through informal interviews with program developers, to deliver insights at a 'framework' level (Coleman et al., 2017; Heeter et al. 2018). At a finer, more granular level than these approaches, Moleka et al. (2022) gathered project-specific data from program administrators willing to share data on their project's actual metrics/drivers of financial performance such as costs, Internal Rate of Return (IRR), tax treatments, renewable energy credits (RECs), and revenues. The data also included current project financial models such as how long it took to secure financing, the scalability of the financing approaches, and model performance. The research found that the four most significant barriers to LMI community solar development were "project/portfolio size," "risk mitigation," "host site negotiations including navigating affordable housing regulations," and "time needed to structure and execute the deal."

5 Conclusions and Recommendations

Community solar projects serving LMI households are one of the fastest growing segments of the solar industry (Greco and Archbell 2022b). The LIFT project researched optimal LMI community solar project finance models and customer experience models across three utility service environments – IOUs, MOUs, and ECs. By examining the experiences of subscribers of community solar programs, LIFT research found that it is important to engage customers at the start of their sustainability journey. The results also revealed that most existing community solar capacity serving LMI households, is deployed in states with enabling legislation – where subscriber savings are twice that of states without enabling legislation (see Table 1).

In terms of recommendations, the LIFT research revealed that real or perceived risks to customer acquisition and subscriber management can be overcome to allow developers to realize competitive financial returns and customer retention for this important market segment. This can be done through the Inflation Reduction Act (IRA). The IRA further underscores the importance of the LIFT recommendations in that in the post-IRA market context, easier access to tax credits and thus project finance value stacking, solar developers can significantly reduce their operating costs, thereby delivering higher bill savings to LMI subscribers. As a result, industry leaders and advocates must keep driving the policy changes that will maintain the accelerated growth and increased savings of LMI community solar. Unfortunately, to-date, relationships with financiers have not caught up with this expectation and persistence is needed for LMI-inclusion in community solar project design (Moleka et al. 2022; Greco and Archbell 2022a). Similarly, more regulations on energy efficiency programs, including cost-effective financing mechanisms are needed to help improve renters' experience in multifamily housing (Reina and Kontokosta 2017; Chai et al. 2021; Hernandez 2015). The LIFT dataset is the only community solar LMI-specific dataset in existence – implying that enhancing the database and adding new functionalities to the LIFT Toolkit are the main avenues for future research.

References

Bohr, J., McCreery, A.C.: Do energy burdens contribute to economic poverty in the United States? A panel analysis. Soc. Forces **99**(1), 155–177 (2020)

Brown, M.A., Soni, A., Lapsa, M.V., Southworth, K., Cox, M.: High energy burden and low-income energy affordability: conclusions from a literature review. Prog. Energy **2**(4), 042003 (2020)

Buylova, A.: Spotlight on energy efficiency in Oregon: investigating dynamics between energy use and socio-demographic characteristics in spatial modeling of residential energy consumption. Energy Policy **140**, 111439 (2020)

Coleman, K., Blank, T.K., Probst, C., Waller, J.: Financing community-scale solar, January 2017. RMI

Chai, A., Ratnasiri, S., Wagner, L.: The impact of rising energy prices on energy poverty in Queensland: a microsimulation exercise. Econ. Anal. Policy **71**, 57–72 (2021)

Elnakat, A., Gomez, J.D., Booth, N.: A zip code study of socioeconomic, demographic, and household gendered influence on the residential energy sector. Energy Rep. **2**, 21–27 (2016)

Greco, V., Archbell, J.: Research Report: customer experience for low-and moderate-income community solar subscribers, June 2022a. Elevate

Greco, V., Archbell, J.: Research Report: The growth of U.S. community solar serving low-and moderate-income households, June 2022b. Elevate

Heeter, J., Bird, L., O'Shaughnessy, E., Koebrich, S.: Design and implementation of community solar programs for low and moderate-income customers, December 2018. NREL

Heeter, J., Xu, K., Chan, G.: Equitable access to community solar: program design and subscription considerations, July 2021. NREL

Heindl, P.: Dynamic properties of energy affordability measures. Energy Policy **86**(c), 123–132 (2015)

Hernandez, D.: Sacrifice along the energy continuum: a call for energy justice. Environ. Justice **8**(4), 151–156 (2015)

Hernández, D., Bird, S.: Energy burden and the need for integrated low-income housing and energy policy. Poverty Public Policy **2**(4), 5–25 (2010)

Hernández, D., Phillips, D.: Benefit or burden? Perceptions of energy efficiency efforts among low-income housing residents in New York City. Energy Res. Soc. Sci. **8**, 52–59 (2015)

Hilbert, A., Werner, M.: Turn up the heat! Contesting energy poverty in Buffalo, NY. Geoforum **74**, 222–232 (2016)

Kontokosta, C., Reina, V.J., Bonczak, B.: Energy cost burdens for low-income and minority households. J. Am. Plann. Assoc. **86**(1), 89–105 (2019)

Liddell, C., Morris, C.: Fuel poverty and human health: a review of recent evidence. Energy Policy **38**(6), 2987–2997 (2010)

Liddell, C., Morris, C., McKenzie, S., Rae, G.: Measuring and monitoring fuel poverty in the UK: national and regional perspectives. Energy Policy **49**, 27–32 (2012)

Li, K., Lloyd, B., Liang, X.-J., Wei, Y.-M.: Energy poor or fuel poor: what are the differences? Energy Policy **68**, 476–481 (2014)

Moleka, E., Bacharach, C., Robichaux, E.: Research Report: Analysis of solar project finance research, June 2022. Groundswell

Moleka, E.: Energy impoverishment and Energy Insecurity in the United States, September 2021. Groundswell

Mohr, T.M.: Fuel poverty in the US: evidence using the 2009 residential energy consumption survey. Energy Econ. **74**, 360–369 (2018)

Ray, A., Wang, R., Nguyen, D., Martinez, J., Taylor, N., Searcy, J.K.: Household energy costs and the housing choice voucher program: do utility allowances pay the bills? Hous. Policy Debate **29**(4), 607–626 (2019)

Reames, T.G., Daley, D.M., Pierce, J.C.: Exploring the nexus of energy burden, social capital, and environmental quality in shaping health in US Counties. Int. J. Environ. Res. Public Health **18**(2), 620 (2021)

Reina, V.J., Kontokosta, C.: Low hanging fruit? Regulations and energy efficiency in subsidized multifamily housing. Energy Policy **106**, 505–513 (2017)

The Shelton Group, SEPA: What the community solar customer wants? (2015). https://sheltongrp.com/insights/what-the-community-solar-customer-wants/. Accessed 18 May 2023

Wang, Q., Kwan, M.-P., Fan, J., Lin, J.: Racial disparities in energy poverty in the United States. Renew. Sustain. Energy Rev. **137**, 110620 (2021)

Wright, F.: Old and cold: older people and policies failing to address fuel poverty. Soc. Policy Adm. **38**(5), 488–503 (2004)

Governing Solar Benefits as a Transparent, Local Common Good: The Solar Commons Dashboard

Kathryn Milun(✉)

University of Minnesota, Minnesota Design Center, 84 Church Street, Minneapolis, MN 55455, USA
kmilun@d.umn.edu

Abstract. Over the coming decade of energy transition, solar PV will generate billions of dollars in "solar savings" on the electric bills of whomever owns the solar array. Once these savings have paid for the costs of the system, they are "free" and abundant gifts of the sun. What legal, economic, and educational tools do we need to equitably distribute these solar savings as revenue streams governed by and for local communities doing reparative justice work? The Solar Commons Project (SCP) is prototyping tools to own solar assets and benefits as community common wealth trusts. The toolkit includes legal templates and a digital dashboard that provides real time information to all parties of the trust agreement. The dashboard shows the flows of value from the sun's radiance to kilowatt hours, market-valued solar savings, social wealth trust funds and community impacts. The dashboard is an essential tool of the Solar Commons ownership model. This article describes the dashboard being prototyped by the author for three Solar Commons projects in Arizona and Minnesota. It discusses the new economy vision and the essential components that make this dashboard a tool for local, transparent peer governance of solar energy as a common good.

Keywords: Commons · Reparative Justice · Community Solar

1 Introduction

Alongside a dynamic social-ecological landscape, the value and benefits of solar energy technology have changed dramatically over the past seven decades. Solar made an early, whimsical appearance in the 1950s at the beginning of what social scientists call the Great Acceleration. This was a time when the US stepped into the industrial production gap left in the ruins of post-World War Two Europe. America's successful assist in rebuilding Europe's industrial infrastructure and expanding its own rested on its vast supply of fossil fuel energy resources. The Great Acceleration names how we now look back on that time as the beginning of an American-lead acceleration of socio-economic trends that include vertiginous GDP growth and voracious increase in the use of primary energy and water resources. The famous "hockey stick" charts correlate that growth with planetary increases in air and water pollution, loss of forests, ocean acidification and rise in earth temperature (See Fig. 1).

D. Renné et al. (Eds.): ASES SOLAR 2023, SPE, pp. 178–184, 2023.
https://doi.org/10.1007/978-3-031-39147-7_20

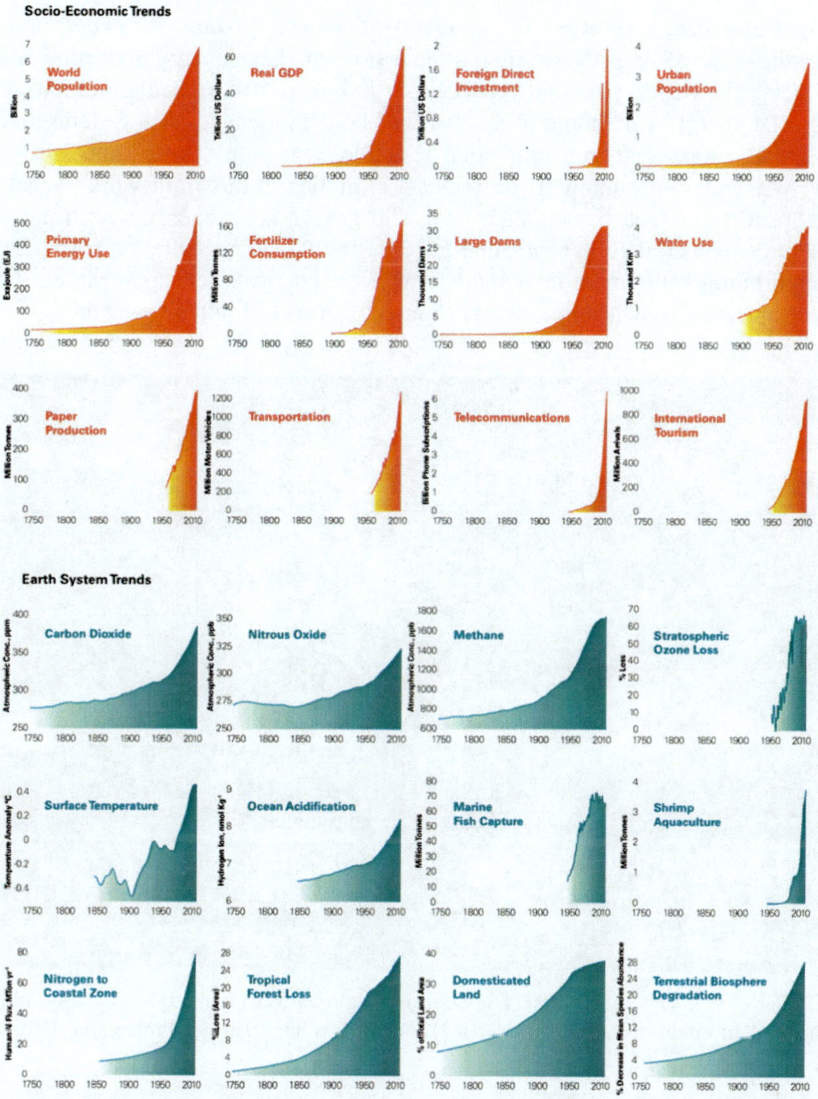

Fig. 1. The Great Acceleration of human influence on the planet across biogeophysical and social measures. Source: Adapted from Steffen, W. et al. 2015. The Anthropocene Review, 2(1), 81–89.

During this period of "the American Century," a solar cell-powered device was just an advertising gimmick--the "Solar Do-Nothing Machine" appeared in 1954 as part of an advertising initiative of the Aluminum Company of America (see Fig. 2). At a cost of $76US/Watt, solar did not figure among the "cheap" and seemingly inexhaustible energy resources that powered the Great Acceleration [1]. But by 2021, with the cost of utility-scale PV at $1.14/Watt, solar is among the cheapest sources of electric power [2]. While we celebrate this great techno-economic achievement of the solar industry, we should also pause to reflect on what this all means for how we understand and measure the value of solar today. Should the history of cheap energy that culminates in the US being the largest per capita energy user in human history be the framework for valuing solar energy? [3] Should we be worried about unintended earth systems impacts that might accompany approaching solar technology as both "cheap" and "limitless" energy? What socio-economic assumptions are implied in this framework? What else are we apt to reproduce if we value, own, and govern solar energy within the same 20th century industrial framework that gave us the Great Acceleration? And finally, what opportunities might we miss by valuing solar energy as cheap, clean electricity powering an economy whose success is framed in terms of limitless growth?

Fig. 2. Charles and Ray Eames et al. The Solar Do-Nothing Machine, 1957. Courtesy John and Marilyn Neuhart cited in Barber, D. A. (2013). The World Solar Energy Project, ca. 1954. Grey Room, (51), 64–93. (p. 66)

As we respond to the urgency of the climate catastrophe with massive solar deployment, solar advocates are also seeking to answer these questions. US social justice activists note that the industrial scale roll out of solar energy within the dominant growth economics of monopoly investor-owned electric utilities (IOUs) reproduces economic inequities [4]. There are unnecessary delays and obstacles as IOUs resist smaller scale rooftop and community scale solar for which they are not the owners [5]. Immense, capital-intensive solar farms are causing harm to soil hydrology and displacing prime farmland around the country [6]. Like the sacrifice zones that surround fossil fuel extraction sites, the impacts of industrial scale solar are often considered necessary to get the "efficiencies" of quick and cheap solar deployment. Whether such solar deployments are carried out by public or private entities, viewing the value of solar through the eyes of industrial scale players and owners brings a legacy of assumptions that many solar advocates claim are not warranted by the distributive nature of the technology itself [7]. The US Department of Energy is aware of these inequities and concerns and is offering new incentives for experimentation with "community solar" ownership and expansion of what we mean when we talk about the value and "benefits" of solar [8].

In this short essay, I discuss a community solar project that addresses many of the concerns outlined above. The Solar Commons Project, which I founded and direct, is a community-engaged experiment with a new (and ancient) way of owning, governing and, in the end, viewing solar technology and its benefits. We use strategies of community trust ownership which allow a solar host to own the solar array while the solar savings on the host's electric bill are owned by an underserved community group (See Fig. 3). Our first, small (14.5 kW) Solar Commons prototype was interconnected to the grid in Tucson, AZ in 2018. It was awarded a US DoE Sunny Award for Equitable Community Solar in 2023. We have two other 500 kW prototypes under development in rural and urban Minnesota. After describing the basic elements of our community trust solar ownership model, I focus on the digital dashboard tool we have built to visualize the flows of wealth in the solar energy system. This is a peer-governance dashboard tool used by all parties to a Solar Commons Trust agreement. We are testing the dashboard over the next two years with community partners in our "living lab" prototype sites. By discussing the design and use of this digital dashboard tool, I can best demonstrate how solar technology, when owned as a civic sector "commons," can generate new kinds of equitable solar benefits and, importantly, can make visible a paradigm shift from the twentieth century view of cheap energy as an engine for limitless growth to a twenty-first century view of energy as abundant common wealth and as a vehicle for regenerative, local wealth-building.

Fig. 3. Infographic of Solar Commons Community Trust Ownership Model.

2 Commons and Solar Commons

Commons are an enduring social form at hand for shaping the socio-economic potential of solar energy. In the English language and the experience of English folks, commons refer to ancient and persistent forms of local community economy. From collective agricultural fields to shared forests for food, energy and housing needs, commons involve customary (informal) social rules of access, just portion, and stewardship. Historically, commons have been diminished or disappeared as modern forms of power divided nature's common wealth into either public or private property [9]. Over the last centuries, natural resource extraction occurred at ever larger scales requiring capital investments of large private entities or sovereign states. The build out of massive twentieth century fossil fuel and nuclear energy infrastructures are part of this history of the disappearance local energy commons. But the distributive nature of the photovoltaic technology that emerged in the 1950s along with the popularity of the "appropriate technology" movement that arose in the 1970s made community scale energy ownership experiments popular in industrialized countries and helped to reawaken the social form of commons [10].

There are many features of solar technology that lend itself to the vocabulary of commons. The sun shines for everyone. Photovoltaics capture the sun's common wealth and turn it into clean electricity. To become a solar commons, the common wealth benefits—which need not include the actual electricity—need to be locally governed and equitably distributed. The Solar Commons™ ownership model does this through a twenty-five-year Solar Commons Trust Agreement using industry standards to calculate a building host's solar savings and a trust-based method to share those savings with a community beneficiary. The agreement is subject to a Creative Commons License establishing social equity and reparative justice as general guiding principles for using the trust funds [11]. When the Rocky Mountain Institute analyzed the Solar Commons ownership model in 2018 using data from Arizona, Minnesota, and Colorado, it found that current net-metering rules created net present positive financial value when not scaled beyond five hundred kilowatts. Larger scales would add administrative costs and risks that would take money away from the trust beneficiary [12, 13]. Around this locally manageable size and equitably guided agreement, Solar Commons researchers have built one other commons tool to create transparency and accountability for all parties to the trust: a digital flow-of-values dashboard.

3 The Solar Commons Dashboard

At its 500 kW scale, we understand that a Solar Commons system in Minnesota could make approximately $70,000 a year for its trust beneficiary. But this sum only represents the market value of the solar savings flowing annually over the twenty-five-year lifespan of a solar array under a Solar Commons Agreement. But if we begin to count the value of the stability of the revenue stream flowing over time to the underserved neighborhood, or the common good that the community beneficiary builds in their neighborhood with the trust funds, we need a different way of accounting for the values and benefits of solar energy, one which sees solar technology embedded in ecological, technological, and social relationships that generate a flow of diverse values. To see this, the Solar Commons team created an interactive "Flow of Values" digital dashboard tool that is used by all parties to a specific Solar Commons Trust Agreement—solar host, trustees, beneficiaries and trust protectors. Below is our prototype dashboard for the 14.5 kW Solar Commons host in Tucson (See Fig. 4).

The dashboard makes visible, in real time, information about the sun's changing radiance (orange); data from the Tucson host's solar inverter and the solar savings seen on host's monthly electric bill (yellow). The dashboard sets those values into a relationship with the ecological value of averting C02 emissions (blue), the volatile market value/price of electricity (dark green), and the stable social values created by the SC Trust Agreement which include the accumulated SC trust funds (light green) and the common good created by the neighborhood beneficiary (pink) who upload photos to report their community work. In Tucson, kids from an elementary school in the trust beneficiary neighborhood are distributing the annual Solar Commons funds using a school "participatory budgeting" process that brings radical democracy and community decision-making into the social form of twenty-first century commoning.

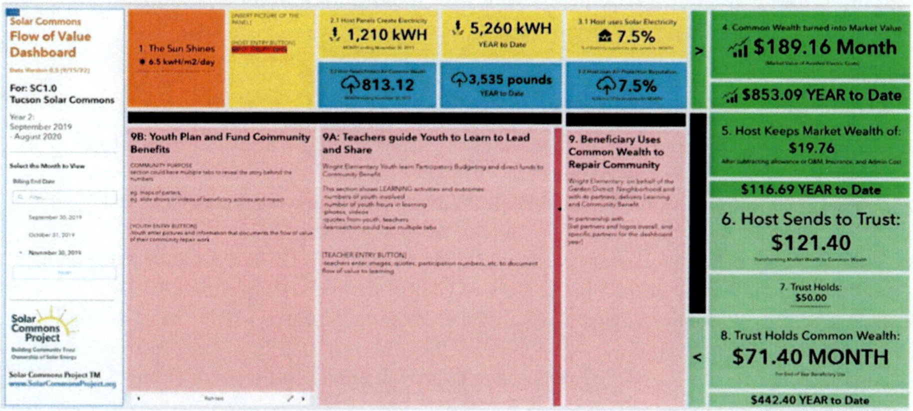

Fig. 4. The interactive Solar Commons Dashboard Tool showing real time flow of values from the sun's radiance to kWhs of solar electricity generation to market value and the to social trust value and finally to common good in the beneficiary neighborhood.

Over time Solar Commons researchers are gathering data and monitoring good governance for how communities use the Solar Commons Dashboard. Our belief is that the dashboard will create the transparency and accountability needed by all parties to the SC Trust Agreement. As we test and improve the dashboard tool with community feedback, we will be increasing the trustworthiness of our Solar Commons legal and peer-governance toolkit and coming closer to co-creating Solar Commons as a robust institution of a commons economic sector. What are the benefits of solar that appear using a Solar Commons ownership model? They will be as diverse as the local solutions that the myriad Solar Commons beneficiaries come up with to build and govern community wealth in their neighborhoods.

4 Conclusion: Shaping Solar Technology for Livelihoods that Value the Abundance and Regenerative Capacity of Nature

The Solar Commons dashboard is part of a commons sector governance structure that visualizes the work of nature, the work of technology, and the mutual aid work of community members. Over time, the dashboard is part of what builds social trust. The Solar Commons dashboard centers the value of solar energy in common wealth: the shared common wealth of the earth's physical and social systems. The market value of solar is but one among many forms of value that the sun's common wealth passes through. Importantly, tools like the Solar Commons dashboard make visible a paradigm shift from the modern twentieth century view of cheap energy as an engine for extractive economies fashioned for limitless growth to a twenty-first century view of energy as abundant common wealth and a vehicle for resilient, diverse forms of locally governed community wealth.

In 2023, solar energy is no longer a background toy to the Great Acceleration. But it is also more than an industrial tool to stop climate pollution. The value of solar includes its great potential to be governed to power a more equitable, resilient future.

References

1. History and the Development of Photovoltaics. AE-Solar (n.d.). https://ae-solar.com/history-of-solar-module/. Accessed 5 May 2023
2. Annual Technology Baseline: Utility-Scale PV. National Renewable Energy Lab (n.d.). https://atb.nrel.gov/electricity/2022/utility-scale_pv. Accessed 5 May 2023
3. Patel, R., Moore, J.W.: A History of the World in Seven Cheap Things: A Guide to Capitalism, Nature, and the Future of the Planet. Univ of California Press, California (2017)
4. Baker, S.: Revolutionary Power: An Activist's Guide to the Energy Transition. Island Press, Washington, D.C. (2021)
5. Burke, M.J., Melgar, R.: SDG 7 requires post-growth energy sufficiency. Front. Sustain., 89 (2022)
6. Yavari, R., Zaliwciw, D., Cibin, R., McPhillips, L.: Minimizing environmental im-pacts of solar farms: a review of current science on landscape hydrology and guidance on storm-water management. Environ. Res. Infrastruct. Sustain. **2**(3), 032002 (2022)
7. Gao, X., Zhou, S.: Solar adoption inequality in the US: trend, magnitude, and so-lar justice policies. Energy Policy **169**, 113163 (2022)
8. Department of Energy: General Guidance for Justice40 Implementation. US Department of Energy, 1 July 2022. https://www.energy.gov/sites/default/files/2022-07/Final%20DOE%20Justice40%20General%20Guidance%20072522.pdf. Accessed 5 May 2023
9. Capra, F., Mattei, U.: The Ecology of Law: Toward a Legal System in Tune With Nature and Community. Berrett-Koehler Publishers, Oackland (2015)
10. Pursell, C.: The rise and fall of the appropriate technology movement in the United States, 1965–1985. Technol. Cult. **34**(3), 629–637 (1993)
11. Milun, K., Walsh, T., Pitner, M.: "Bringing New Light to One of The Oldest Forms Of Property Ownership: An Innovative Solution for Benefiting Underprivileged and Under-served Communities Using the Solar Commons Community Trust Model." (forthcoming 2023, Vermont Law Review)
12. Brehm, K., Lillis, G.: Solar Commons Financial Analysis Results: Solar Commons Project Analysis Phase 1 of 2 Rocky Mountain Institute (2018a). https://solarcommonsproject.org/wp-content/uploads/2020/12/RMI_SolarComFinancialReportPhase1_.pdf
13. Brehm, K., Lillis, G.: Solar Commons Scalability and Constraints Analysis: Solar Commons Project Analysis Phase 2 of 2 Rocky Mountain Institute (2018b). https://solarcommonsproject.org/wp-content/uploads/2020/12/RMI_SolarComScalabilityReportPhase2_.pdf

Priorities' Role in Community Solar: Survey-Based Study and Payment Performance Analysis

Zahra Thani[1], Winslow Lewis[1], Liz Neyens[2], and Jacob Ford[3](✉)

[1] Energy Allies, New York City, USA
[2] Boston, MA, USA
[3] Solstice Power Technologies, Cambridge, MA, USA
jake@solstice.us

Abstract. This study aims to comprehensively analyze community solar project preferences for consumers and suppliers by conducting three distinct analyses. First we analyze the predictors of community solar contract adoption to understand how individual priorities affect the probability of adoption. Using an original data set of survey responses from potential community solar customers, we analyzed the predictors of contract adoption by employing a weighted logit model. We find that individuals who were previously familiar with community solar projects were significantly more likely to adopt than those who were not familiar. Secondly, a survey of community solar developers and financiers identified industry perceived barriers to community solar access and inclusion. Thirdly, we gathered payment performance information from community solar initiatives to measure how financial risks are perceived and how they interact with customer demographics. Our study contributes to the literature by providing insights into the drivers and barriers of community solar adoption and sheds light on the importance of understanding individual priorities in designing effective community solar policies.

Keywords: Community solar · contract adoption · individual priorities · income-qualified · inclusion

1 Introduction

In 2020, the United States solar industry installed a record 19.2 gigawatts of solar capacity, representing a 43% increase compared to the previous year [1]. As solar expands at a rapid pace, concerns arise about the fairness and inclusivity of renewable energy access. Often, there are disparities in who receives benefits from renewable energy [2]. For example, the conventional method of installing solar panels on rooftops of homes limits the benefits to individuals who have homes with rooftop conditions that can support panels, limiting access for renters or households with insufficient direct sunlight [3].

L. Neyens—Independent Researcher.

D. Renné et al. (Eds.): ASES SOLAR 2023, SPE, pp. 185–195, 2023.
https://doi.org/10.1007/978-3-031-39147-7_21

Community shared solar (CSS) presents a solution to increase the distribution of the benefits of renewable energy. Previous research found CSS increases the market size for customers and shares costs, alleviating financial and locational barriers to solar participation [3, 4]. Because community solar projects are typically at an off-site array, they do not require individuals to own a home with a specific type of roof. Instead, individuals who are renters, lack appropriate roof conditions, or are unable to finance a solar installation, can subscribe to or buy into a portion of the community solar program.

Community solar ownership models including special purpose entities (SPE), utility-sponsored, and non-profits, offer opportunities to expand renewable energy benefits [8]. In the SPE model, a separate legal entity outside of the utility is created to develop and manage the community solar project, while utility-sponsored models involve the utility itself initiating and managing the project. The non-profit models are driven by community organizations or entities operating for the public benefit. Future adoption of the respective model depends on factors like policy support and regulatory restrictions. To date, twenty-one states support CSS through legislation, with states like New York and Colorado embracing various ownership models, with SPE and utility-sponsored being most prominent [9].

CSS participants receive a credit on their utility bill, resulting in a discount on their total energy cost, and communities that participate may receive other benefits including workforce development, investment into community infrastructure, and a reduction of fossil fuels. As a result, CSS offers a potential avenue for more low-to-moderate income (LMI) individuals, also known as income-qualified individuals, to access clean energy. A National Renewable Energy Laboratory (NREL) report estimates that as of the end of 2020 only 65 MW, or 2%, of that total capacity is dedicated to income-qualified subscribers [5].

Existing research in CSS ownership models and customer preferences notes the lack of a comprehensive analysis of preferences for both consumers and suppliers, particularly for disadvantaged populations [10]. To address this gap, this study aims to fill the research void by conducting three distinct analyses. Firstly, it examines the sensitivity of potential customers to CSS contract attributes and develops quantitative measurements to gauge individual priorities that may impact the adoption of contracts. Secondly, it surveys CSS developers and financiers to qualitatively identify the most significant factors that limit the offering of CSS to income-qualified customer bases. Finally, the study analyzes the monthly payment performance data of two operating community solar farms to determine the demographic or socioeconomic factors that are correlated with the likelihood of churn. By integrating the findings of these analyses, the study aims to provide a more comprehensive understanding of CSS preferences for consumers and suppliers, as well as identify ways to increase adoption and analyze churn rates among income-qualified individuals.

2 Methods

This research incorporates data from three distinct datasets: (1) Community Solar Priorities Research: survey forms completed by respondents of varying demographics assessing the favorability of factorial test contracts and offering feedback on various aspects of community solar product; (2) Developer and Financier Research: survey responses from solar developers and financiers to assess price sensitivities of language used in contract agreements; and (3) Churn and Default Research: records of subscribers leaving programs and missing payments provided by solar developers. Additional information about survey methodologies, copies of survey questions, and regression equations can be found on the public GitHub pages made available by the authors [11].

2.1 Community Solar Priorities Survey

To evaluate potential customers' willingness to sign up for CSS and identify contract attributes that are correlated with differing rates of contract adoption, a survey was conducted in which respondents were asked to evaluate two hypothetical community solar contracts with varying contract terms. Additional feedback in the form of open response questions was collected on various solar product offerings but is not reported for brevity. Demographic data was self-reported from respondents. To analyze likelihood of contract adoption, a logit model is used to estimate respondents' stated willingness to enroll in the contracts they reviewed, controlling for available demographic characteristics.

The survey data was collected in 2021, resulting in 1,493 unique individuals and 2,986 contract reviews, as each individual reviewed two contracts. We sought to produce a representative sample, developing a sampling scheme and survey weights using data from the American Community Survey (ACS). Additional information is available from the authors on the sampling scheme and weighting methodology, and is omitted here for space constraints.

2.2 Developer and Financier Survey

The developer and financier research consists of an analysis of survey data in which each respondent was asked to provide estimated community solar contract attributes, including estimated savings rate, term length and cancellation fee. Respondents were asked if these rates differ when being offered to income-qualified respondents, and if the difference was positive or negative relative to the general population. Information on potential loan loss reserve characteristics, including ideal length of time and total project revenue to be covered, was also collected.

Additionally, respondents reported the relative degree of importance of community solar access for income-qualified individuals is qualitatively coded using a five-point Likert scale. These factorized survey questions included three main sections. The first related to a variety of factors that may influence the respondent's ability to make community solar projects accessible to income-qualified populations. Secondly, which internal factors were most important that lead to the inclusion of income-qualified populations. Finally, what factors affect the project's ability to fill out enrollment in a timely and efficient manner. For brevity, we report only selected findings.

The developer and financier survey data were collected in 2022 resulting in 254 responses received. Due to the small sample size and exploratory nature of this research, we simply report the topline results of this survey without further analysis or significance testing.

2.3 Churn and Default

Customers in existing community solar projects typically pay a monthly subscription fee. Participation in community solar farms is voluntary, hence customers may exit the program at any point, occasionally subject to a cancellation fee determined in the contract. Churn refers to the act of a customer exiting the CSS program whereas default refers to the customer failing to pay the monthly subscription fee.

Monthly account-level data was collected from two community solar projects from January 2020 to April 2022, resulting in 812 accounts. We supplemented account-level information with individual-level demographic data from Experian, including Experian's Income Insight Score (a proxy for income), Experian's VantageScore (an alternative credit score metric), and other demographic data such as gender, education, occupation, and homeownership status.

3 Results

3.1 Community Solar Priorities Survey

A logistic regression model is employed to measure Y_{signup}, the probability of contract adoption, controlling for the following contract attributes and demographic variables:

$$
\begin{aligned}
Y_{signup} = {} & \beta_0 + \beta_1 x_{savings\ rate} + \beta_2 x_{contract\ years} + \beta_3 x_{contract\ pages} \\
& + \beta_4 x_{cancel\ fee} + \beta_5 x_{income} + \beta_6 x_{race} + \beta_7 x_{homeowner} \\
& + \beta_8 x_{familiarity} + \beta_9 x_{contract\ reviewal\ time} + \epsilon_i
\end{aligned}
$$

A logistic regression model shows the probability of contract adoption, with independent variables including a number of contract attribute and demographic variables. Contract attributes include savings rate, contract length and pages, cancellation fees, time spent reviewing the contract and of which contracts were reviewed. Demographics include income, race, homeownership status, and familiarity with community solar.

Table 1 below shows the contract attributes and demographic data from the surveyed population, along with the rates of contract adoption ("Signup Status"). The contracts that respondents reviewed varied along the following features: (1) the program's savings rate compared to the typical monthly electricity bill, (2) the program's cancellation fee, (3) the contract term's length in years, and (4) the page length of the contract.

Table 2 details the probability of contract adoption across various contract terms. When holding all other variables constant, low-income individuals are less likely to adopt contracts compared to high-income individuals. From the second model results, the odds of contract adoption for medium income participants are 1.8 times those of low-income participants. No statistically significant difference in contract adoption is observed when respondents are shown differing savings rates.

The strongest indicator of probability of sign up is familiarity with the subject of community solar. Relative to less informed participants, more informed participants are 4.4 times more likely to sign up, holding constant all other contract and demographic attributes.

Table 1. Summary Statistics by Signup Status

Signup Status	No, N = 1,215	Yes, N = 1,771
Savings Rate		
5% savings rate	418 (34%)	583 (33%)
10% savings rate	404 (33%)	599 (34%)
20% savings rate	393 (32%)	589 (33%)
Cancellation Fee		
$250 Cancellation Fee	612 (50%)	879 (50%)
Zero Cancellation Fee	603 (50%)	892 (50%)
Contract Length Pages		
10-page contract	602 (50%)	889 (50%)
20-page contract	613 (50%)	882 (50%)
Contract Length Years		
1-year contract	572 (47%)	924 (52%)
25-year contract	643 (53%)	847 (48%)
Race		
01 White	511 (42%)	693 (39%)
02 Black	223 (18%)	357 (20%)
03 Hispanic	151 (12%)	263 (15%)
04 Asian	192 (16%)	242 (14%)
05 Other POC	138 (11%)	216 (12%)
Homeowner Status		
01 Homeowner	652 (54%)	1,076 (61%)
02 Renter	563 (46%)	685 (39%)
Income		
01 High	333 (27%)	439 (25%)
02 Low	398 (33%)	420 (24%)
03 Moderate	484 (40%)	912 (51%)
Familiar with CSS		
01 Less familiar	804 (66%)	550 (31%)
02 More familiar	411 (34%)	1,221 (69%)

3.2 Developer and Financier Survey

The results from the Developer and Financier Survey are presented in Figs. 1 and 2. Figure 1 shows the factors for making CSS accessible to income-qualified customers. 55.5% of developer respondents highlighted customer acquisition costs (CAC) as the largest barrier to enrolling income-qualified customers, more than risk of subscriber

Table 2. Probability of Contract Adoption Regression Results, Odds Ratios

Variable	Probability of Contract Adoption Odds Ratios						
10% savings rate	1.131	1.071	1.062	1.124	1.125	1.17	1.17
20% savings rate	1.098	1.071	1.072	1.08	1.081	1.124	1.127
02 Moderate income		1.433**	1.367*	1.354*	1.360*	1.195	1.193
03 Low income		0.801	0.742*	0.794	0.793	0.700*	0.706*
02 Black			1.184	1.183	1.178	1.095	1.078
03 Hispanic			1.31	1.367	1.378	1.185	1.175
04 Asian			0.904	0.93	0.932	0.985	1.005
05 Other POC			1.225	1.461	1.458	1.555*	1.558*
02 Renter				0.730*	0.722**	0.913	0.933
20-page contract					0.938	0.936	0.945
25-year contract					0.801*	0.796*	0.799*
$250 Cancellation Fee					1.013	1.039	1.043
02 More familiar						4.355***	4.388***
02 More review							1.291*

Reference Categories: Savings Rate - 5%; Income - High Income; Race - White; Homeownership - Home Owner; Contract Pages - 10 pages; Contract Years - 1 Year; Cancellation Fee - Zero; Familiarity - Less Familiar; Review Time - Less

*p < 0.05 **p < 0.01 ***p < 0.001

churn, risk of subscriber default, lack of project financing, difficulty qualifying income-qualified households, and difficulty communicating with income-qualified households.

Figure 2 shows the top factors that influence the decision if developer and financiers will include LMI customers in their projects. The importance of policy requirements is key when deciding whether to include LMI individuals in community solar projects. 63% of developers noted that policy requirements are very or extremely important. This was the highest-rated category compared to company interest, developer/financier interest, equity and inclusion, and community interest.

3.3 Churn and Default

For the final component of the research, we analyze probability of churn or default in two active community solar farms. A logistic regression model is employed to measure Y_{churn}, the probability of churn. Logarithmic transformations for continuous variables are used, namely VantageScore and Income Insight Score.

$$
\begin{aligned}
Y_{signup} &= \beta_0 + \beta_1 x_{Tenure} + \beta_2 x_{Solar\,kWh} + \beta_3 x_{Gender} + \beta_4 x_{VantageScore} \\
&+ \beta_5 x_{IncomeInsight} + \beta_6 x_{Occupation} + \beta_7 x_{homeowner} \\
&+ \beta_8 x_{Marital\ Status} + \beta_9 x_{Education} + \epsilon_i
\end{aligned}
$$

Table 3. Churn and Default Summary Statistics

Variable	**Overall**, N = 812	**Farm A**, N = 454	**Farm B**, N = 358
Churn	118	59	59
Default	6	1	5
Average Tenure (Months)	22	22	23
Average kWh	6.1	5.9	6.3
Gender			
Female	182	117	65
Male	193	111	82
Marital Status			
Married	228	134	94
Single	33	27	6
Occupation			
Healthcare/Education Services	25	16	9
Management/Technical	6	0	6
Self Employed/Other	23	14	9
Services	14	6	8
Education			
Completed College	99	61	38
Graduate Degree	114	63	51
High School or Some College	150	95	55
Income Insight Score, Average	122	114	167
Average VantageScore, Average	804	803	806

Note that given the few observations of default, results are only shown for churn events. Table 3 shows the descriptive statistics for the demographic and socioeconomic factors of the CSS customers (Table 3).

A logistic regression determines the probability that an account will experience churn, with results in Table 4. Increased tenure length is associated with decreased probability of churn. This effect is measured even when controlling for a combination of demographic and socioeconomic variables. For example, a one-month increase in tenure is associated with a 0.15 decrease in log odds of churn.

4 Discussion

This research provides insight into barriers to accessing community solar. The community-facing survey offers insight into how community solar programs can incorporate the priorities of climate-impacted groups such as income-qualified individuals.

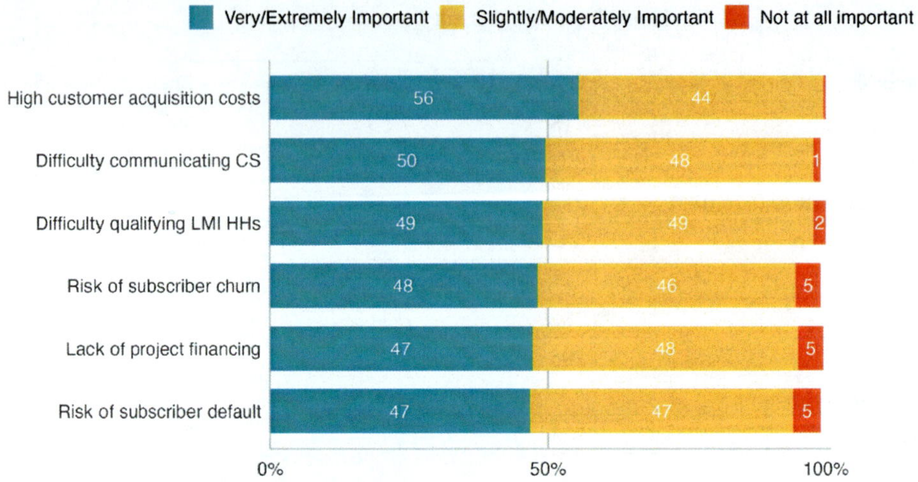

Fig. 1. Factors to make CSS Accessible to LMI Customers

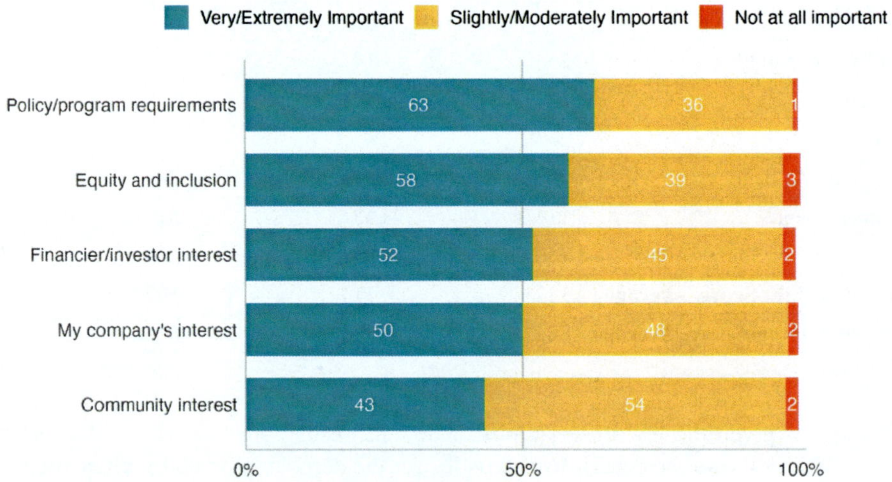

Fig. 2. Factors to influence decision to include LMI Customers in Projects

The developer and financier survey helps reveal what the industry perceives as barriers to climate-impacted communities' inclusion in their programs. Finally, the analysis of payment performance data from active solar projects points to little relationship between demographic and socioeconomic factors and the observed risk of churn.

Our community-facing survey reveals key takeaways about individual preferences in community solar programs. Our data shows that an increased savings rate may not result in more income-qualified participation. Instead, in our data, familiarity with community

Table 4. Regression Results, Probability of Churn

Variable					
Tenure	−0.195***	−0.141***	−0.152***	−0.150***	−0.185***
kWh Solar	−0.045	−0.008		−0.025	
Gender (Female)		−0.554*			−0.019
Vantage Score Logged	0.164	0.069			−0.083
Occupation					
Management/Technical					0.067
Self Employed					0.191
Services					−0.657
Homeowner		0.088	0.246		1.315
Income Logged			0.045	0.068	
Married				0.883	
Education					
Graduate Degree			0.491	0.447	
High School or Some College			−0.201	−0.165	
Other			−1.261	−0.878	

*p < 0.05 **p < 0.01 ***p < 0.001

solar is the strongest factor to influence likelihood of contract adoption, demonstrating the importance of education initiatives when encouraging sign-ups. Shorter contract lengths also influence likelihood of contract adoption among our respondents - individuals preferred 1-year contracts over 25-year contracts.

Our developer-facing survey points toward key next steps for organizations who advocate for climate-impacted groups. Across all reasons for including and prioritizing climate-impacted groups in community solar projects, developers are most likely to ensure inclusion when policies require it. States that require a certain percentage of project capacity to flow to climate-impacted communities are critical to meeting President Joe Biden's Justice40 Initiative, whose goal is to deliver 40% of federal investments in clean energy to low-income communities and communities of color. Developer respondents also highlighted customer acquisition costs as a barrier to including income-qualified individuals. Customer acquisition costs for income-qualified populations are higher compared those of the total population [6]. More research is needed in this area to increase access to CSS. For example, many CSS programs require excessive documentation for income eligibility determination. This trend demonstrates a need for a more streamlined system. Moreover, data transparency could help lower overall projects costs by ensuring that project stages such as interconnection happen in a timely manner.

Lastly, our analysis of payment performance data challenges the notion that income-qualified individuals pose a higher risk to the success of community solar projects.

Our findings found no statistically significant correlation between income level and the likelihood of defaulting on payment.

Our results demonstrate the benefits of education for communities on community solar starting from project inception. Individuals who are familiar with the potential benefits are more likely to sign up for a CSS program. Additionally, policies requiring income-qualified carve outs in CSS projects are important to ensuring that Justice40 initiatives are met, as developers noted that they are most likely to include income-qualified individuals when legislation makes it a requirement. Support from nonprofit partners such as energy justice organizations that advocate for these policy changes are critical to spur inclusion.

Previous research has demonstrated that various groups are likely to be left out of renewable energy projects. Yet, to the best of our knowledge, previous research has not collected either survey data on community solar contract preferences or payment performance data from actual community solar projects.

As a project that was the first of its kind for the community solar industry, this study faced a number of challenges. Regarding the community-facing survey, we were limited by measurable variables. Our demographic variables were self-reported, and many of the trends present in our data are first-time findings that require future research to validate. The survey weighting methodology required aggregating a number of racial groups into an "other person of color" category, limiting our ability to include a more disaggregated analysis. For the analysis of payment performance data, the probability model was limited due to a low sample size of churn events.

Future research could be helpful in creating a broader understanding of community solar preferences across the United States. Additionally, more work on the correlation between policy change and inclusion is a critical next step in understanding how barriers to access could be impacted.

Ultimately, community solar contract terms can be improved for communities. Individuals have preferences that are not being met by current offerings. However, income-qualified inclusion in the clean energy transition goes beyond contract terms. Developers should prioritize community input at every project stage. Policies are instrumental to ensuring inclusion, and creating measurable requirements for climate-impacted inclusion in community solar projects is critical to ensuring that the benefits for clean energy projects flow to climate impacted communities.

References

1. International Energy Agency. Renewables 2020: Analysis and Forecast to 2025
2. Carley, S., Konisky, D.: The justice and equity implications of the clean energy transition. Nature Energy **5**, 569–577 (2020). https://doi.org/10.1038/s41560-020-0641-6
3. Nuñez-Jimenez, A., Mehta, P., Griego, D.: Let It Grow: how community solar policy can increase PV adoption in cities. Energy Policy **175**, 113477 (2023)
4. National Renewable Energy Laboratory: Shared Solar: Current Landscape, Market Potential. Impact of Federal Securities Regulation. https://www.nrel.gov/docs/fy15osti/63892.pdf
5. National Renewable Energy Laboratory: What is Community Solar 2: Cost Impacts to Participants and Non-Participants. https://pubs.naruc.org/pub/C82FE238-1866-DAAC-99FB-4B78C72230AB

6. Klein, S., Hargreaves, A., Coffey, S.: A financial benefit-cost analysis of different community solar approaches in the Northeastern US. Solar Energy **213**, 225–245 (2021)
7. National Renewable Energy Laboratory: Design and Implementation of Community Solar Programs for Low and Moderate-Income Customers. https://www.nrel.gov/docs/fy19osti/71652.pdf
8. Coughlin, J., et al.: Guide to Community Solar: Utility, Private, and Non-Profit Project Development, 1–56 (2011). https://digitalscholarship.unlv.edu/renew_pubs/39
9. Heeter, J., Xu, K., Chan, G.: Sharing the Sun: Community Solar Deployment, Subscription Savings, and Energy Burden Reduction. National Renewable Energy Laboratory (2021). https://www.nrel.gov/docs/fy21osti/80246.pdf
10. Klein, S.J., Hargreaves, A., Coffey, S.: A financial benefit-cost analysis of different community solar approaches in the Northeastern US. Solar Energy **213**, 225–245 (2021)
11. Ford, J.: SETO Data Analysis GitHub Repository. https://jake-ford.github.io/SETO_Data_Analysis/

Author Index

D. Renné et al. (Eds.): ASES SOLAR 2023, SPE, pp. 197–198, 2023.
https://doi.org/10.1007/978-3-031-39147-7